The Bicycle in Wartime
An Illustrated History
Revised Edition

THE BICYCLE IN WARTIME

An Illustrated History

Revised Edition

Jim Fitzpatrick

Star Hill Studio

This revised edition published in Australia in 2011
by Star Hill Studio Pty Ltd
ABN 12 135 353 249
342 Mt. Kilcoy Road,
Kilcoy, Queensland, 4515, Australia
www.starhillstudio.com.au

Copyright © Star Hill Studio Pty Ltd 2011

All rights reserved.
No part of this publication may be reproduced,
stored in a retrieval system, or transmitted,
in any form or by any means without
the permission of the publisher.

ISBN: 978-0-9807480-1-7

National Library of Australia Cataloguing-in-Publication entry
    Author: Fitzpatrick, Jim, 1943-
    Title: The bicycle in wartime: an illustrated history/Jim Fitzpatrick.
    Edition: Rev. ed.
    ISBN: 9780980748017 (pbk.)
    Notes: Includes bibliographical references and index.
    Subjects: Military cycling--History.
              Bicycles--History.
              Cycling--History.
    Dewey Number: 623.74

First Edition published by Brassey's, Washington and London, 1998

Designed by Roey Fitzpatrick

Contents

Preface	vi
Chapter 1. Bicycling—The Experimental Years	1
Chapter 2. The Boer War	51
Chapter 3. The War of the Words: 1900–1914	77
Chapter 4. World War I: The Allied View	91
Chapter 5. Die Radfahrtruppe: 1914–1945	111
Chapter 6. When Tojo Came A-Wheeling	141
Chapter 7. The Home Front: World War II	157
Chapter 8. The Bicycle in Vietnam	175
Chapter 9. Retrospect and Prospect	205
Notes	215
Sources of Illustrations	236
Acknowledgements	240
Bibliography	243
Index	258

Preface

The Bicycle in Wartime is a history of the use of bicycles and bicycle technology in wartime, and is concerned with three underlying elements.

The key concept is that of bicycling, which involves using bicycles, but not necessarily riding them. A soldier with a bicycle is not the same as a soldier on a bicycle. There is an immense difference between the two, in what each can—and can't—do.

A second theme is how the technology has been exploited and developed for military purposes. The bicycle revolutionized the human transport scene in the 1890s. It was lightweight, strong, durable, reliable, needed little maintenance, and no fuel, food or water. The energy efficient cyclist proved much faster than a horse, and could carry 100 pounds or more (plus the rider). Not surprisingly, many a military mind took note.

Of special interest is how the bicycle's use and potential have been perceived. Many military minds over the years have dismissed it as irrelevant or obsolete for warfare, while others adopted it to great effect against their enemy, such as the British in the Boer War, or the Japanese in their invasion of Malaya-Singapore, to mention only a couple. Unfortunately, few writing about those events have properly acknowledged the cyclists' role, if at all. Perceptions can be narrow and memories short.

This revised version of *The Bicycle in Wartime* includes updated opening and final chapters. Only a few new photos have been included since in the internet world there are now many visual and written records of the cycle's wartime use and equipment readily accessible, and increasing almost daily. This book provides a framework for putting it all in perspective.

Jim Fitzpatrick
Kilcoy, Queensland, Australia
April, 2011

1. Bicycling—The Experimental Years

The warfare of this decade is quite a scientific study.
　　　　　　　　　　　　THE AUSTRALIAN CYCLIST, FEBRUARY 6, 1896

The cyclist soldier has the advantage of the cavalryman in superiority of speed, and noiselessness of travel, without a cloud of dust to betray him, ease of concealment, and without the necessity of forage for his mount.
　　　　　　　LIEUTENANT WILLIAM T. MAY, UNITED STATES ARMY, 1892

The Commander-in-Chief himself [Lord Wolseley] says that as long as he can ride a horse, he will not learn to ride a bicycle.
　　　　　　　　　　　　　　　THE SKETCH, SEPTEMBER 30, 1896

The concept of *bicycling* is fundamental to understanding the role the machine has played in wartime. A military cyclist is not a soldier on wheels, but a soldier *with* wheels. There is an immense difference, though many never appreciated the fact, or came to grips with the implications. While "riding a bicycle" is the usual image, *bicycling* is essentially a man-machine combination that allows mode to be matched to terrain, optimizing the use of wheel and foot. When sand, mud, obstacles, high winds, or a steep incline make pedaling difficult, the rider can get off and walk. The cycle can be pushed, carried, lifted over fences, and floated across rivers. Heavy weights and bulky loads can be transported on it. Moreover, man and machine can be readily carried on wagons, trucks, cars, boats, or trains. It was that combination that radically altered the human travel equation—for civilians and military alike.[1]

During the 1890s the bicyclist's speed over various distances and varied terrain quickly became apparent, and the machine's carrying capacity demonstrated. Experiments and trials were undertaken in various countries to attempt to determine just how useful and effective two-wheeled soldiers might be. Some were zealous in their efforts to prove the machine's military value, especially bicycling fanatics and manufacturers with vested interests—and there were plenty of both. On the other hand, many refused to consider any evidence placed before them—whether regarding the bicycle or any other of the myriad technological developments that were occurring around the world. Somewhere in between were those simply trying to look at the situation with reasonable objectivity.

The Velocipedes[2]

In early 1818 a German, Baron Karl von Drais, patented the forerunner of the bicycle. It had a wooden frame with two wooden wheels, the front one steerable, and was propelled by the rider pushing his legs against the ground while seated astride. It enjoyed a brief vogue, and the confirmation that a man could balance and steer such a device was significant. However, the *draisienne* was heavy, uncomfortable, inefficient, and had no practical use.

Draisiennes, Paris, 1818

An American Pickering velocipede, 1869

In the 1860s pedals were fitted to the front wheel of a draisienne. This highly efficient technique for using the legs to power a machine was a landmark in technological history. The resulting *velocipede* was produced in significant numbers and the velocipedist could travel faster than a man could walk. However, the wood or iron rimmed wheels proved very uncomfortable on rough surfaces (the nickname "boneshaker" was explicit and appropriate) and for a rider to maintain a 10 mile per hour pace a pedaling rate approaching a hundred revolutions a minute was necessary. The machine was simply not conducive to long, pleasant or fast journeys. The initial craze collapsed around 1870, but it introduced the word *bicycle* into the English language, and saw the creation of a cycle industry. During the subsequent Franco-Prussian war both sides apparently tried velocipedes for dispatch delivery but a scarcity of information makes it difficult to assess the extent or effectiveness of their use.[3]

The problem of the early velocipedes' low speed and rough ride was overcome by greatly increasing the front wheel diameter. That was made possible by the development of light, tensioned wire spokes, first practically applied to bicycles by 1870. By 1875 other innovations included hollow (cushion) rubber tires, and Grout's invention for the accurate tensioning of spokes, which permitted wheel alignment. By 1880 both ball and roller bearings were used and a variety of comfortable saddles developed. Originally the high-wheeled machines were called *bicycles*, but when the "safety" models came in (more in a moment) it became necessary to differentiate between the two styles. As the front wheels became increasingly large, the rear wheels were made smaller, to save weight. Hence, they became known as *high wheelers*, *penny-farthings*, or *ordinaries*.

Serious military experimentation was ushered in by the ordinary bicycle, with its large scale production, widespread adoption, and considerable speed. On them, riders in favorable circumstances managed to cover the mile in under three minutes, 250 miles in one day, and 1,170 miles in six days. Simul-

Albert Pope, manufacturer of Columbia bicycles and military cycling advocate, second from left, at Readville, Massachusetts, 1879

A French artist's impression of a dispatch rider on his "ordinary" bicycle during military exercises

taneously, a great variety of three and four wheeled versions were developed. Harry Griffin's *Bicycles and Tricycles of the Year 1886* catalogues the wide array then available. Unfortunately, the high-wheeled machine was difficult to learn to ride, could be unstable and dangerous, often had ineffective braking, and the solid and cushion rubber tires had a tendency to twist off the rims. Worse, from a military perspective, was the fact that the only place a rider could carry a rifle, ammunition, and pack was on his back—and that made things even more unstable. Other than for messenger service, the ordinary machine was not very practical. However, the tubular metal frames and wire-spoked wheels were adapted to mobile stretchers and wheel chairs, and the relatively lightweight but very strong tricycles and quadricycles set many to thinking about their cargo and weapons carrying capabilities.

Inventors explored various ways of making ordinary bicycles safer and easier to ride. One stream concentrated upon reducing the size of the front wheel, while using gearing systems to keep the gear ratio high. The result was a series of lower, more stable bicycles known collectively as *dwarf* safeties, but which still resembled the high wheelers. A quite different developmental approach led to the modern safety bicycle. In 1885 John K. Starley built a model using the diamond frame concept, with a chain-and-cog rear wheel drive, and a low seating position between two nearly equal sized wheels. Subsequent modifications resulted in the Humber pattern of 1890. Its raked steering fork had ball bearings and was "set" (bent forward at the bottom, an important factor in effective steering and balancing). There was also an adjustable rear wheel for tensioning the chain; light, wire spoked wheels of nearly the same diameter; and a completely triangulated frame, with saddle pillar.

The dwarf safeties and diamond frame models competed briefly for popularity, but the mechanical, structural and riding

Military cycles at the Stanley Exhibition, London, 1888

advantages of the Humber pattern proved far superior. The rider could mount, dismount and sit stationary with no difficulty. The rear wheel drive eliminated the inconvenience of pedaling the steering wheel, and the seating position lessened the incidence of headers. Although the relatively smaller wheels meant a rougher ride, that problem was effectively eliminated with the advent of pneumatic tires in 1888. By 1890 the diamond frame dominated the cycling world, and that basic design has been used ever since for the vast majority of bicycles.

The Humber safety bicycle design of 1890

The Safety Revolution[4]

The two things that most caught the attention of civilians and military alike were the speed and strength of the safety bicycle. Undoubtedly the most spectacular aspect was the speed. The seeming epitome was reached in the United States, when "mile-a-minute" Murphy pedaled 60 miles an hour behind a train in 1899. Of far more importance, however, was the fact that bicycling enabled the average person to negotiate the countryside faster than by any other means of personal transport, including the venerable horse. By 1898 a cyclist had ridden 428 miles in only 24 hours. For the average rider, 75 miles in a day was readily achievable, and 100 miles or more not at all unusual. Equally amazing was the ability of cyclists to maintain those speeds indefinitely, even under difficult conditions. In Australia, Pat O'Dea averaged 103 miles daily during his 1,700 mile ride across the arid, isolated Nullarbor Plain. There was no formal road over part of the distance, and much of the surface of the few existing roads was sand. No horse could remotely match such speeds over such distances.

A particular source of admiration and wonder was the machine's strength and carrying capacity, despite its light weight (models in the 1890s commonly weighed between 24 and 29 pounds). A popular advertisement of the era showed one brand supporting 16 men with an aggregate weight of 2,448 pounds. It was no gimmick. In engineering tests diamond frames have supported 4,000 - 5,000 pounds before permanently deforming or collapsing. The strength of the light weight frame and wheels was quickly put to practical use. Australia's thousands of sheep shearers, in the course of their annual labor migrations about the sparsely settled continent, routinely loaded their bicycles with 75 pounds of luggage and water, plus the rider's weight. Importantly, everything could be mounted on the bicycle itself, greatly lessening rider fatigue. Indeed, a commonly acknowledged giveaway of an inexperienced bush cyclist was the sight of a pack carried on the back.

Aside from the diamond frame and pneumatic tire, the other significant development was the freewheel hub. The first safeties were fixed wheel. That is, the pedal and rear wheel always

One of many such advertisements in the 1890s touting the strength of the bicycle

turned when the bicycle was rolling. If a rider wanted to coast rapidly downhill, for example, he had either to lift his legs and let the pedals spin under him, or pedal at a high rate in order for his feet to keep up with the rapidly rotating crank and pedals. As well, fixed wheel machines often had no separate brake (the rider slowed by resisting the forward motion of the pedals). Some freewheel hubs were fitted to production bicycles from about 1897, caught on fast, and were increasingly common from the turn of the century. Getting on and off and coasting were much simpler, and they made multiple gear systems practical. In particular, the patenting of the Sturmey-Archer three-speed hub in 1902, and the two-speed Eadie hub, with backpedal or coaster brake, in 1903, made many cyclists' lives much easier and safer.

The safety bicycle was sound, serviceable, durable, and required relatively little maintenance. That is well illustrated by an Australian Post Office assessment of its bicycle fleet, in 1940. Many bicycles in outback towns had not been into central workshops for two decades, had uncertain maintenance histories, and administrators became concerned as to how safe they might be. Twenty of the oldest known machines, ranging up to 23 years in service, were collected from around the country. All had been used six days a week, most for two deliveries a day, carrying up to 50 pounds (plus rider) per delivery. Several were estimated to have

Typical bicycle load of an Australian rural worker, 1911

Tricycle cart, 1897

covered more than 100,000 miles. Just two were junked (and one of them had been in daily use until removed for the study) and only minor maintenance was required for the others.[5]

During the course of the worldwide bicycling craze, which reached its height in 1895-1896, national cycle industries expanded rapidly and began to saturate local markets. As American, British, Canadian, French, and German manufacturers cut prices drastically and sought overseas outlets, bicycles became progressively cheaper. By 1900 a new one could be had for the equivalent of two or three weeks' wages in some countries, a used one for less. For those not having the cash, "hire purchase" or "time payments" were popular, and numerous workers managed to scrape up the monthly bicycle payment. In fact, the cycle was an important factor in influencing the development and widespread public acceptance of the time payment concept in several countries.[6]

Tyres and Terrain[7]

After John Boyd Dunlop's invention of the pneumatic tire and tube combination in 1888, several key advances rapidly followed. One was the beaded-edge tire, with one or more strands of wire wound around the bead. As well, wheel rims were narrowed and deepened. The result was that the beaded tire was held firmly on the rim by the inflated tube pressure, yet was relatively easy to remove for repairs when deflated. In 1892 came the diagonally threaded fabric lining, which allowed shock absorption without undue distortion of the tire cover. Together, these developments led to the rapid, widespread, practical use of the pneumatic tire. It was ultimately the key to effective travel over diverse surfaces, and "did more to popularize cycling than did the cycle itself."[8]

Leaks posed difficulties initially, but with improved tube and valve joins, better rubber compounds, and thicker casings they became less common. The major problem cyclists faced was punctures from sharp stones, thorns, nails and glass. By the late 1890s Dunlop, reacting to feedback from Australian and South African cyclists, had marketed a "Thorn Proof" tire which featured twice the fabric thickness and 50 percent more tread than usual. While it increased rolling resistance, its puncture resistant capabilities were legendary, and many felt that the extra pedaling effort was well worth it. However, most cyclists did not use Thorn Proofs and still traveled widely with little difficulty. Indeed, a myth that must be laid to rest is that the tire at the turn of the century was of poor quality. The bicycle's use in Australia, for example, demonstrated that by the late 1890s the tires were extremely durable even in harsh riding conditions.

The nature of bicycling meant that pedalers applied different standards to riding surfaces. While the pneumatic tire was effective at providing a smooth ride and avoiding energy loss for small-scale roughness (about three-quarters of an inch or less), larger obstacles were another matter. Cyclists—who could maintain from 10 to 20 miles per hour, and attain 40 miles per hour downhill—found that rough surfaces, potholes, ruts, limbs and rocks were not merely uncomfortable or inconvenient, they were dangerous. As well, every time the bicycle bounced up and down it meant a direct loss of energy, which the rider had to personally supply.

Even if riding conditions were not always as desired, however, the cyclist was rarely worse off than anyone else. More often, he was far better off, because the great advantage of the pneumatic-tired machine was that it allowed the rider not only to use existing roads, whether ideal or not, but to negotiate many off-road conditions. The cyclist could ride over surfaces that disintegrated under the weight of heavier vehicles or the sharp hooves of animals. Alternatively, if necessary, the rider could

The Australian Dunlop Thorn-Proof tyre, illustrated at bottom right

push or carry the bicycle until riding was again feasible. And with the ability to lift the machine over fences and logs, and to utilize narrow footpaths, game trails and camel and sheep pads, the cyclist found opportunities for rapid travel that were denied other modes of transport. The bicyclist quickly proved to be

A bicycle pad in the remote West Australian outback, 1935

faster and more flexible across more varied terrain and surfaces than any other transport mode of the time, carrying considerable cargo in the process. While good roads certainly made for more comfortable riding, the machine's real serviceability and advantage lay in its versatility.

In the final analysis, the bicycle radically enhanced human mobility. Travelers' perceptions of time, distance, and rate of movement in 1890 were very different from today, and can probably be fully appreciated only by those who were totally and inescapably confined within that framework. Suddenly, they were presented with the safety bicycle. With it, the effort formerly required to walk 20 miles could now get a traveler up to one 100 miles (depending upon surfaces and terrain)—and the rider's possessions could be strapped on the machine instead of being carried on the back. From that perspective, bicycling was a truly revolutionary development. It is no wonder that so many sat up and took note of its military potential.

The Cavalry and the Steel Steed

By the mid-1880s several armies had experimented with cycling. Authorities in Austria, in 1884, and in Germany, in 1886, reportedly bought or tried bicycles. The French army formally had introduced them into service by mid-1887. In England, the first recorded military cycle use was during Easter Manoeuvres in 1885, with cyclists serving as scouts. At least three Volunteer Corps cyclist sections were formed that year. At the next Easter Manoeuvres a movement of bicycle infantry was tried *en masse*. The earliest British military unit consisting solely of cyclists was the 26th Middlesex (Cyclist) Volunteer Corps, established in February 1888.[9]

The Connecticut National Guard Signal Corps was formed in 1891. It was the first formal military bicycle unit in America. It was soon followed by the Second Company of the District of Columbia National Guard, under General Albert Ordway. Both were actively supported by the Pope Manufacturing Company, which was owned by Albert Pope, a Civil War veteran and maker of the popular Columbia bicycle. Units in Illinois and Colorado were created not long after. They variously paraded, delivered messages, experimented with the machine in local conditions, and were reported upon by newspapers and cycle journals.[10]

In June 1888, at the Royal United Service Institution, Lieutenant Colonel A. R. Savile, a member of the Cyclist Touring Club and Professor of Tactics at the Royal Military College, presented a lecture on military cycling. *The Saturday Review* account of that meeting noted that one attendee, Colonel Lonsdale Hale, offered "adverse criticism" when the merits of cyclists *vis a vis* cavalry were raised by Savile. Lord Wolseley, who chaired the meeting, defended Savile when the "combative qualities" of Colonel Hale "carried him rather further in his remarks than he intended." The cyclist versus cavalry die was cast.[11]

Lieutenant Colonel A.R. Savile

The rider-machine combination was commonly equated to the cavalryman in the early days. Several military cycling advocates brought the comparison upon themselves by emulating horsemen; they tried tilting at rings with lances as they pedaled past, or cutting off a simulated head, mounted on a stake, with swords. The latter proved far more difficult than expected. While a horseman can swing about in the saddle "with impunity," as one reported, cyclists trying to swing a sword with great force were thrown off balance. Observers quickly concluded that the bicyclist did not have the shock capacity of a charging cavalryman, and lacked "the moral effect of the flashing of the sabres" in quelling riots and dispersing mobs. Few disputed that observation. Among the last things a cyclist wants is to run into anything, or to try swinging a sword at an armed, nimble opponent while simultaneously pedaling and steering a machine across a rough surface. From that perspective, the bicyclist was certainly no contender for a cavalry role. It led an American officer to conclude that the military cyclist was "nothing more than an infantryman, who must get off his steel horse to fight."[12]

Many, however, appreciated that a bicyclist is not a horseman, and should not try to be one. This was highlighted at military

The First Signal Corps of the Connecticut National Guard, America's first formal military bicycle unit, 1891

cycling trials at Aldershot, on October 7, 1887. There, according to the Earl of Albemarle, a cyclist on a cavalry obstacle course dismounted at each obstacle, lifted his machine over, climbed back on, and pedaled off. There was considerable debate over whether it was fair.[13]

A number of cycling aficionados contended that the cyclist was actually superior to the horseman. They usually framed their arguments around the mechanical nature of the device. One common comparison was that the machine was less likely to be hit or put out of action by enemy fire than a horse. That certainly is true, in that there is much less to a bicycle, and substantial parts of it could withstand a hit without affecting its running. As one writer commented, a bullet hitting a cycle "only requires the gunsmith's aid," but a horse "cannot be carried to the repair shop." Expanding upon that theme, an Englishman added that even if several bicycles are damaged, new ones can be assembled from the interchangeable parts, "a power of resurrection which the most bigoted antipathist must credit is not possessed by the cavalryman." An Australian suggested that "a low set, swiftly moving cyclist, swinging along the winding track, would be a more difficult object to aim at than the horseman. Also dozens of bullets might pass through the machine without touching it or at least without doing much damage." However, as an American countered, once a cyclist *is* hit, he does not have his trusty horse to get him home.[14]

In 1895 a German suggested that machine guns and magazine rifles were making cavalry obsolete, advocated a reduction in their number, and urged a greater employment of cycle scouts. Not many soldiers yet accepted that view. Most still saw the bicyclist as inferior to the horseman and, at best, as only a "substitute" in certain circumstances. Even comparisons favorable to the bicycle were often couched in horse terms: a one-ounce oil can would supply forage for months; it needed limited stabling facilities; it did not neigh to betray its position; its grooming was of the most elementary nature; it was the lacquered mount.[15]

While it was years before some sorted out how cyclists differed from equine troops, a number quickly appreciated the inherent and complementary value of the bicyclist. General Grandin, a French cavalry officer, believed that "cavalry and bicyclists

"Tent pegging" and "tilting at the ring"

should work together as part of the same corps, and not as rivals." Charles Turner, in America, likewise felt that the issue was not what were the advantages of the bicycle over the cavalry, or vice versa, but how well pedalers and horsemen could aid one another.[16]

Ultimately the bicyclist and horseman were two separate entities with different performance capabilities and roles to play. Many, however, never saw, nor wanted to see, beyond the horse-cyclist debate. The comparison was perhaps inevitable, but unfortunate, for it obscured the more fundamental issue of exploring the innovative nature of cycling and cycle technology, and assessing how best to use the bicyclist to maximum advantage.

Messengers and Relay Riders

The military initially adopted the bicycle because it proved to be a "cheap, effective and speedy accessory" for delivering dispatches. By the mid-1890s several armies used cycle couriers and orderlies in cavalry, artillery, infantry, and medical corps. The dispatch spirit seemed irrepressible. A Connecticut National Guard cyclist demonstrated that he alone could deliver a message faster than an entire flag signaler network. A cyclist relay team carried a dispatch 975 miles—from General Nelson A. Miles' Chicago headquarters to that of General Howard, in New York City—in 4 days and 13 hours, in mostly rainy weather. In Germany, soldiers pedaled in relay from Munich to the North Sea in 81 hours. And in Australia, a team of Melbourne riders raced a 30-word message several miles from the city to the port, and returned with a reply, quicker than a team of semaphore signalers could do it. While they proved the point, no military decision maker down under saw any sense in replacing six two-flagged men with six two-wheeled men, for the sake of six minutes.[17]

In 1894, the ante was upped when a bicycle relay team carried a message 2,000 miles, from Washington, D.C. to Denver, in six and a quarter days, "one of the most notable feats on record in transmitting information by human power alone, over the greatest space in the shortest time." Two years later an even longer relay ride was undertaken between San Francisco and New York, known as the *Journal-Examiner*-Yellow Fever Transcontinental Relay Ride, after the newspapers on the east and west coast which sponsored it. "The greatest thing of its kind ever undertaken," the 3,400 mile relay required three months organization by two men (each making two cross country trips), extensive use of the telegraph and railway networks to establish and coordinate the relay legs, and 220 riders pedaling sectors averaging 15.5 miles.[18]

In the Rocky Mountains and deserts, where settlements were scarce, riders were provisioned at isolated posts. The cyclists rode over 8,000-foot passes, through storms, and a third of the way at night. The overall average was 11 miles per hour, but along some stretches they managed 20 miles per hour. The dispatch was handed over at the *Journal* offices, before a crowd of several thousand. A ceremonious final leg was pedaled/paddled on a flotation water-bicycle from the Battery to Governor's Island, and a colossal, illuminated bicycle parade held that night. The ride, at the peak of the worldwide cycling craze, convinced an English reporter that "military cycling has achieved its greatest feat in America."[19]

As impressive as the great relay rides were, they ultimately proved of little interest to military communications planners. Their organization took months and required the use of the very telegraph and railway systems that they were meant to replace in an emergency. In the final analysis, the bicycle courier's greatest value was not as an operative in a fixed network, but in being able to ride 100 meters or 100 miles at a moment's notice, with no support. The bicyclist's forte lay in being a highly flexible, mobile element within a military organism which, in wartime, could find itself in great flux.

Mounted Infantry

It was impossible to discount the potential of bicycling infantry—they simply traveled too fast. Even with heavily loaded machines, "those not especially expert" could cover 10 miles an hour for hours on end without fatigue. And the machine was not just a fair weather mount. As one officer observed, "Supposing the cyclist cannot [ride] on account of temporary bad roads. What of it? Are you worse off than before?" No. He need only dismount and push the machine. In the eastern United States, a Lieutenant in the Ninth Infantry traveled over 100 miles per day for three days with his loaded cycle. An officer from the Seventeenth Infantry bicycled 530 miles in five and a half days, carrying his rifle, revolver, rations, blanket, canteen, dog tent, and 30 rounds of ammunition. And in England, after a group of soldiers had pedaled 175 miles in two days, General Sir Frederick Maurice admitted that there was "no cavalry in the world which can touch that." Those interested in the rapid movement of ground troops could not ignore such achievements.[20]

It was seemingly a field commander's dream: "Think of a force of infantry going in almost any direction without the assistance of rail or boat, 40 or 50 miles in one morning and appearing on a field not leg weary and exhausted as from two days' hard march, but comparatively fresh and ready to fight all the afternoon!" As well, when cycle-mounted infantry dismounted to fight, they would not have to leave one of every three or four riders behind to hold bridles, as did horse-mounted soldiers, nor would temporarily abandoned bicycles on the ground be as visible as horses, or as likely to be damaged en masse by artillery fire. And as for enemy scouts, they would not be able to tell from the cycle tracks which way the troops were going. The logic and advantages seemed irresistible to some.[21]

But there were some difficult issues to be resolved. For one thing, cyclists traveling with horsemen, wagons and infantry could far outdistance them. Consequently, the speed of the cyclist unit (a major reason for having them in the first place) was sacrificed. That led many to conclude that bicycle troops ought to operate relatively independently: "A hundred wheels or less can be handled easily and be made a powerful factor in warfare." However, they could not be handled as easily as some suggested. Officers noted the "natural desire" of some individuals to race, the indulgence of which was "fatal to military discipline." As well, the tendency for a cycle column to "tail out" meant that 100 cyclists single file on a narrow road stretched for half a mile. It was impossible to issue immediate verbal orders in such circumstances. On the positive side, one writer felt that cyclists would make excellent escorts for baggage trains, as they would not be "tempted to mount the wagons," as an infantry escort sometimes is. On the other hand, in a retreat, "especially if precipitous, disorganized bodies of cyclists would be a nuisance."[22]

Finally, First Lieutenant R. G. Hill, of the 20th U. S. Infantry, applied his military mind to another problem:

> Suppose that the head of the [cyclist] column arrives at the foot of a slope steep enough to make it desirable to dismount. If the distance of six feet is to be preserved throughout, the entire column must take up the speed of the head, and the speed of the rear is sacrificed. The head of the column arrives at the top and has a level stretch ahead, but the rear must walk to the top of the hill, and, if distances are to be preserved, the speed of the head will be sacrificed. The entire column has walked from the time the head arrives at the hill until the rear leaves it.[23]

With several centuries of experience at moving armies of men, horses, wagons and cannon, and having worked out rates of travel and sequences of movements, the military was suddenly faced with a radically faster element that would require a lot of thinking through. It was no wonder that many in the military balked at pressing such a device into service, and hence into the service manual.

The ability of cycle troops to cover great distances in a hurry was explored through a series of long overland rides in remote areas and under harsh conditions. They demonstrated both the capabilities of the machine and problems associated with its use. One such journey, commanded by Lieutenant De R. C. Cabell, took place in the isolated country east of Fort Meade, South Dakota, on the extreme eastern edge of the Black Hills. On the cold morning of October 1, 1895, his troops set out on a ride across arid country with no running streams and only a few alkaline water holes. While there were no high mountains, there was "not a stretch of level ground five miles long" between the two army posts.[24]

The Overlanders

Notwithstanding the conditions, Lieutenant De Cabell's men covered 444 miles in just over seven days. He concluded that cyclists could travel over such country, carrying everything necessary, at much greater speed and with less effort and cost than horsemen, who also would have to cart water supplies. During the trip the Lieutenant and his men also experimented with the West African kola nut:

> We used it two days and the effect on each of us was that, while chewing, we felt no fatigue, no thirst, and no hunger. There was not the slightest feeling of intoxication, nor were there any symptoms of reaction after using it ... we had a very light breakfast at 2 A.M., and nothing more till 6 P.M., and I had not felt hungry ... That it would be of the greatest value in long, hard marches, where water and food are scarce, there seems no doubt

The route included extensive stretches of black "gumbo" soil, which when wet stuck to the tires and forks and left the cycles unridable. When dry—which it was when de Cabell's troops passed through—the soil was rock hard, "seamed in all directions with deep cracks," and was barely passable for bicycles as a result of large clods and deep tracks formed during wet weather by horses and wagons. The soldiers were forced to pedal and push along the badly rutted trail, because bunch grass and cacti prevented them from riding on the smoother ground alongside. Although the soldiers were careful, their tires quickly filled with thorns. Surprisingly, they found that as long as the thorns remained in the tires and tubes, they continued to hold air. However, when a thorn worked its way out or, after a couple of days, began to decay, the tires would start leaking.

In an annual report in 1896, General Nelson A. Miles formally recommended the organization of an army cycle regiment. The challenge was taken up that year in Fort Missoula, Montana by a young West Point graduate, Lieutenant James A. Moss. He initiated a series of experimental rides with his recently formed 25th United States Infantry Bicycle Corps, comprised of African-American soldiers.[25]

Members of the 25th Infantry Bicycle Corps

The initial training runs ranged up to 40 miles a day, and included practice in scaling nine foot fences and fording streams with fully loaded bicycles. They accomplished the latter by placing a pole between two men's shoulders, laying the bicycle on its side on the pole, and balancing it with their hands. With all gear left on the machine, they managed to cross in up to three feet of swiftly flowing water. It was no mean feat. The average weight of a loaded bicycle was 80 pounds, with a knapsack, blanket roll, and tent tied to the handlebars; a rifle on the horizontal bar and a sack of food and haversack hanging beneath it; and food and cooking utensils in an 11-gallon tin case mounted on a separate frame on the front of the cycle. Each rider also carried a canteen and cartridge belt on the body.

On their first long ride, in the mountains about Lake McDonald, they were plagued by nearly continuous wind and rain: "Had the Devil himself conspired against us we would have had but little more to contend with." Nonetheless, they managed 126 miles in 24 riding hours. In the notorious gumbo they spent a half hour cleaning off their tires, and on one six-mile stretch were forced to dismount some 20 times because of bad riding conditions and fallen trees in the road. Their shoes filled with mud, their feet often came off the pedals, and there were several falls, especially when riding down steep, slippery slopes. In the process they confirmed several facts about the use of bicycles in mountainous country. Whereas they took a half hour to push their machines two miles up a mountain pass, they rode one and three-quarter miles down the other side in only five minutes. Also, the brakes that Moss had fitted (they were not standard on fixed wheel cycles at that time) were found to be absolutely essential. Under the difficult conditions, the heavy man-machine combinations (average weight, 237 pounds) could not have been controlled without them.

Moss and his men were undaunted. The following week they set out on a 797 mile ride—through rain, mud, and snow—across the Continental Divide near Mullan Pass. It was impressive. Despite atrocious conditions, they negotiated the distance in 126 riding hours. On their worst day they managed only 40 miles in 10 hours; on their best they covered 72 miles in eight and three-quarter hours. Back at Fort Missoula, Lieutenant Moss' troops carried routine messages, mail and telegrams to and from wagon trains. Rides of less than 45 miles a day did not even tire them.

In 1897, with the formal permission of the Secretary of War, Lieutenant Moss led the most publicized military cycle ride of the decade. With 21 others he traveled 2,000 miles from Fort Missoula to Jefferson Barracks, in St. Louis, in 40 days. The journey was reported around the world and was the subject of articles in the *Army and Navy Journal*, *Army and Navy Register*, and *Scientific American*. Lieutenant Moss and the surgeon were white. The other 20 cyclists were African-Americans.[26]

A. G. Spalding supplied bicycles with heavy duty frames, luggage carriers, frame bags, repair kits, and spare parts. They had reinforced forks and crowns to withstand the heavy loads placed on the front. The wheels were of the heavier tandem gauge and had more spokes than normal. Eight different makes of tires were tested on the trip. Geared at 68 inches, each bicycle weighed 32 pounds unladen. Specially designed frying and baking pans which clipped together in pairs were fitted within the diamond frame, and doubled as food containers. Large coffee pots tied on the handlebars had rolled blankets stored inside them. The remainder of the equipment and provisions was loaded as usual, except that rifles were carried on the soldiers' backs. By the end of the journey Moss concluded that nothing should be carried by the rider, and recommended that in future, carbines be substituted and tied along the crossbar. The total weight of machine, rider and luggage ranged from 205 to 272 pounds. Moss received his new bicycles just 10 days before departing for St. Louis, leaving little time to fit out the machines and train. Notably, several new recruits were poor cyclists, and one only learned to ride that week.

The 25th Infantry Bicycle Corps left Fort Missoula on June 14, 1897. Lieutenant Moss kept a meticulous diary. They headed southeast towards Yellowstone Park, where they toured for five days, then on through Wyoming, South Dakota, Nebraska and Missouri. They faced severe weather extremes. On June 17 there was six inches of snow in the Rockies, through which they walked their machines. They crossed the Continental Divide in "an awful sleet storm, with two inches of snow on the ground," and found

Lieutenant James A. Moss and his 25th Infantry Bicycle Corps

it as much work to go downhill in the thawing slush and mud as it had been to push up the other side. A week later, it was 111 degrees Fahrenheit on the plains of Wyoming, and two soldiers suffered blistered feet while pushing their bicycles on hot, sandy surfaces.

Once they left the mountains, they normally rode from daybreak to 10 A. M., then rested until five o'clock, after which they continued until dark in the long, cool summer evenings. Several nights, when they had good roads, they traveled by moonlight. Despite the common image of the great plains as being flat, Moss' men never managed to pedal more than seven miles during the entire journey without having to dismount to push up a slope, hill or mountain. They averaged 52 2/3 miles daily over the 35 riding days (Moss did not count their five days at Yellowstone as part of the trip). The worst single day was near the Platte River, in Nebraska, where heavy rains left nearly impassable conditions, and they negotiated only nine miles. The sandy roads of South Dakota "were awful," as were 200 miles of routes in Nebraska which were eight and 10 inches deep in sand. At one stage, in Montana gumbo country, the road was so bad that they were forced to push their bicycles through adjacent weeds and underbrush. During and after heavy rains they occasionally resorted to using railroad lines. While the surface was solid, the uneven ties and ballast usually made pedaling impossible and they had to push their machines along between the rails.

During his journey, Lieutenant Moss formulated an economic-sociological theory based upon road development:

> Where the roads were properly graded and well worked, the inhabitants were well informed, used modern farming implements, had fine windmills and other conveniences. On the other hand, where the roads were in a bad condition and evidently much neglected, the people were narrow-minded, devoid of any knowledge of the topography of the country, and behind the times in everything.[27]

Moss rated the Missouri roads the worst on the entire trip, and its residents the most "inhospitable."

Moss' men had several periods of short rations until reaching Nebraska, and once rode nearly 50 miles without food or water. For several hundred miles there was only alkali water (they "suffered considerably from its bad effects") and they relied upon railway tanks for potable supplies. As a result, several riders, including Lieutenant Moss, fell ill and were sent ahead by train so that they could recover by the time the others caught up with them. One soldier claimed he was too ill to continue at all, and was sent back to Fort Missoula. However, Moss thought that "as he had given me trouble on several occasions," he was "merely feigning illness." The others finished the journey.

The bicycles held up well. "Although we had quite a number of breakages, frames, spokes, handle bars, balls, etc.," Moss said that they were essentially confined to five bicycles ridden by men who were careless. He thought that more caution on their part would have cut the damage rate at least in half. The front crowns and axles gave the most trouble, not surprising given the extremely heavy loads on the front, and the rough riding they were subjected to. A couple of tires burst or were torn so badly that they were thrown away, but otherwise, Moss did not mention any undue tire troubles. Of the eight makes, one had internal

Moss' men at a river crossing

The 25th along a trail in the Rocky Mountains

springs which allowed it to be ridden even when punctured. However, it lacked resiliency and was unacceptable.

In his report to the War Department, Lt. Moss drew a number of lessons from the journey. Among them was that no more than about 40 or 50 cycle-mounted soldiers could be commanded by one officer, as they tended to string out too far along the road. Also, no fixed rules could be established as to the rate of travel, how long should be allowed between breaks, and so on. The geography of the country determined it, and the commander had to use his judgment. Moss also found that handlebar vibrations incurred through long hours on rough surfaces caused such stress that he and others not only suffered while riding, but were occasionally kept awake at night with pain in the hands and arms. In early 1898 Lieutenant Moss proposed a ride from Fort Missoula to San Francisco, but it did not eventuate. Instead, he soon found himself in Cuba, with a non-cycling unit. His two-wheeled instincts resurfaced in a plan that he drew up for a 100-strong bicycle company to patrol the streets of Havana. He suggested that well armed cyclists could move with "inconceivable rapidity" to riots or wherever else needed. The proposal was not acted upon.[28]

In the midst of the publicity surrounding Lieutenant Moss, wide coverage was also given to Major R. P. Davidson, of the Illinois National Guard, who in late June, 1897, commanded an 849 mile ride to Washington, D. C. Each of his twelve cyclists carried the usual equipment of a soldier in the field, as well as a bicycle tool kit and spare parts. On good days they averaged 76 miles, while on muddy and mountainous roads they managed between 35 and 50 miles. Davidson felt that they could have completed the journey in 10 days, in lieu of the 14 ½ they took, without undue strain. The Major was particularly impressed at how well his troops could push their loaded machines along the worst of the mountainous Appalachian roads without the signs of fatigue that marching men usually displayed carrying gear on their backs.[29]

Cycle Artillery

A number of military units and manufacturers experimented with mounting machine guns and cannons on tricycles. At the English Easter Manoeuvres of 1891, the 26th Middlesex Cyclist Regiment gave the earliest known display by firing a Maxim gun from a cycle carriage. Unfortunately, the unit weighed 96 pounds, had solid rubber tires, and could not be pedaled uphill. Worse, two riders could only manage a few miles an hour even on level ground, and another pair of riders were hooked up to the gun carriage to help pull it along. In 1895 the Pope Manufacturing Company of America fitted a tricycle with a rear-facing small mountain cannon. It was pedaled by two men sitting side by side, and was ridden in a Brooklyn parade. There was no anti-recoil mechanism for the cannon, nor were ammunition cases fitted. The same frame was later used to mount a forward facing machine gun. At the Navy and Military Tournament in London, in 1896, a two-man tricycle with two air-cooled Maxim machine guns was demonstrated by gunners from the 2nd Battalion Tower Hamlet Volunteers. Four bicycle mounted escort riders accompanied them.[30]

The most spectacular cycle artillery, as described in *Scientific American*, was produced by the English company of Vickers Sons & Maxim. It was a large two-man tricycle weapons platform for a pair of machine guns, carrying 1,000 cartridges in metal cases, and two tripods to stabilize it during firing. The total weight was 376 pounds, plus the riders. Although the pair of rear-facing guns could be set up and fired, simultaneously, within a couple of minutes, it was too heavy to pedal up a slope and the riders had to dismount and push. Some observers thought that the unit's effectiveness could have been enhanced, and the weight decreased by 126 pounds, by eliminating one gun, a tripod, and half the ammunition. It would be even better, some suggested, to distribute the guns, ammunition and tripods between three tandem bicycles, which could be pedaled much faster and could more easily negotiate rough surfaces and slopes. Illustrations of the tricycle unit being ridden and fired were widely reprinted in international cycling journals, newspapers, and magazines.[31]

The Vickers Sons and Maxim machine gun tricycle

There were also attempts to mount heavy weapons on standard bicycles. In the United States, Albert Pope fitted a Colt machine gun on one of his cycles. The gun was affixed to the front of the frame by a bracket that allowed it to be rotated 360 degrees. It weighed 40 pounds, excluding ammunition. As well, Pope produced add-ons for his standard tandem model in an effort to interest the military in it. He put clips on the two seat pillars

The machine gun tricycle underway

to hang pistol holsters on; added brackets to hold a matching pair of repeating rifles on either side of the elongated frame; fitted a case with signal flags; and tied army overcoats and blankets on each set of handlebars. His products were displayed at the Madison Square Garden Cycle Show.[32]

Scientific American described an English device in which six twin-wheeled cycle units, each carrying two men, were connected

A Maxim machine gun tricycle at the 1896 Navy and Military Tournament, London

The first known machine gun to be mounted on a cycle carriage, 1891

single file to tow an ammunition vehicle. It reportedly traveled up to 15 miles per hour through London streets, and was able to turn "at great speed" in "a much more circumscribed space than could have been done by an ordinary carriage." The accompanying illustration leaves the viewer baffled, however, as to how the pedal-

Albert Pope's cycle-mounted cannon, 1895

The London Cyclist Battalion's Colt automatic gun

ing mechanism was arranged, and casts doubt upon its ability to turn so tightly.³³

The difficulties inherent in pedaling heavy cycle artillery were ruefully described, in retrospect, in the 1932 history of the London Cyclist Battalion. The editor noted that no reference would be complete without mentioning "the Gun." At one time the Battalion had two machine gun carriages, each pneumatic tired, but with a heavy axle and large, wooden-spoked wheels. No weight was cited, but photographs suggest it was considerable. They were towed by one, two or three cyclists (depending upon the gradient) riding in single file, their bicycles hooked together by bamboo poles. Although "it was the ambition of every hardy cyclist to get posted to the Gun Section," which made "a very pretty show at manoeuvres and in displays and tournaments," it was "only the hardiest who enjoyed it." As well, the resiliency of the pneumatic tires caused the platform to bounce around when the gun was fired, making it difficult to aim.³⁴

In 1896, *Scientific American* felt that "it will not be long before [cycle machine guns] will form a very effective adjunct to regular army service." Five years later the French journal, *La Nature*, predicted that bicycle artillery corps were "destined for a great future." Both were wrong. Not only did their weight make them impractical to pedal, but tricycles have three tracking wheels. If any one of them hits a soft spot, or obstacle, the machine is seriously impeded. As early as 1887, G. Lacy Hillier, who rode many tricycles, concluded that they were essentially unmanageable for general road use. "As for such impracticable absurdities as artillery-carrying tricycles, they need scarcely be mentioned." Time proved him right.[35]

Folding Bicycles

Bicycles can be clumsy to transport on trains and wagons, and awkward to carry for any length of time. The folding bicycle was seen as the answer: "The machine to be carried by man when man cannot be carried by the machine." Indeed, some believed that the development of folding bicycles was fundamental to establishing the cyclist's military niche. It was the subject of much inventive effort in the 1890s, the goal being a machine that could be quickly and compactly folded and unfolded, but which was strong when locked open for normal riding.[36]

The most noted exponent was Captain Gerard, of the 87th Regiment of French Infantry. He experimented with several variations, and in 1895 settled upon a design. He placed the seat over the rear axle, and connected the front wheel to the back by a single frame member incorporating a folding joint. The coupling consisted of two beveled parts linked by a hinge pin. When the machine was in normal riding position, a larger tube on the frame member slid over the connecting joint, overlapped substantially on either side to provide strength, and was locked into place with three screws. The screws, as well as others that adjusted the seat and handlebars, had handles, so no wrenches were required.[37]

With the seat over the rear wheel, the rider could place his feet on the ground and sit stationary while firing or looking around, with no pedals directly below to interfere. As well, he could step onto the bicycle seat from the rear and get underway, or dismount quickly and safely by simply putting his feet down and letting the machine roll out from under him. That was especially advantageous with fixed wheel machines. The disadvantage was that the rider pedaled with his legs forward. A soldier could reportedly dismount, fold it, and put it on his back, using the attached leather carrying straps, in less than a minute, and unsling and remount it even quicker. It weighed 30 pounds, was compact, and "no more cumbersome than a knapsack." Captain Gerard trained 20 men to demonstrate them at French military maneuvers, accompanied by a tandem bicycle carrying an engineer with spare parts.

Gerard's experiments led to changes. He strengthened the frame by increasing the tube diameters and adding a second parallel tube connecting the wheels, effectively duplicating the drop frame structure then used in ladies' models. Each of the parallel tubes used the earlier locking device, but Gerard added an extremely clever ball joint arrangement between them which served as a single pivot point about which to fold the bike. He added a brake block that could be screwed down against the rear tire. It slowed the machine in descent, prevented it from rolling around when not in use, and when barely touching the tire, rubbed nails and other objects off before they could become embedded. Finally, Gerard designed a seat pillar that pivoted about the rear axle, along a curved tube. The rider could adjust the seat through a wide range of riding positions, from the traditional mid-wheel location to one directly over the rear wheel. Gerard's machine, based upon extensive field experience, was highly innovative. It is not clear how many were built or eventually placed in service in the French army.[38]

In contrast to Gerard's approach, most folding bicycle designs were based upon the standard diamond frame model, with

Captain Gerard's second folding model

locking pivots on the crossbar and down tube. One such, built by the Dwyer Folding Bicycle Company, of Danbury, Connecticut, had special handle bars for compact folding, and was clearly aimed at both the civilian and military markets. *Scientific American's* description of the 25 pound machine, modeled by a soldier, noted that the folded unit could be readily "placed in the cabin of a very small yacht," and that a drop frame model was available for ladies. In Austria, *Illustrirte Zeitung* featured a 28 pound folding bicycle based upon the diamond frame, of which 24 were reportedly ridden 2,800 kilometers by a cyclist company during six weeks of maneuvers in Hungary.[39]

Most folding bicycles use some form of hinge-and-lock on the frame tubes. The arrangement must be able to hold the joints absolutely secure, or they will wobble, costing the loss of energy and making the machine awkward to ride. The folding mechanisms often weaken under harsh riding conditions. The inventor of the American Dursley-Pedersen machine tried to overcome the problem by radically altering the basic diamond frame concept. He employed twin thin tubes in lieu of a single thicker one found on the conventional bicycle, in an attempt to increase strength while decreasing weight. The front forks, in particular, were highly unusual. They had four thin tubes running from the steering head (located above the handlebars) to the axle. Where they joined the down tube to the crank bracket, a steel plate held everything in place. To fold the machine, the front wheel unit had to be slipped out of the plate and then pivoted back at the steering head. The steering head also served as one mounting point for the saddle, which was swung hammock-like between a back point. The original model was made from poplar sticks, tied with twine, and the final version constructed with steel tubing. It is not known if production went beyond the prototype.[40]

Of all the specialized military cycling inventions of the era, folding bicycles were one of the few that eventually proved useful. In 1896 a writer for *L'illustration* viewed their potential as analogous to the "fly of the fable," in which the enemy, disdainful at first, would soon be put upon the defensive and finally conquered by the "insect that has tormented it without truce or mercy." With these "flies of war," there would be no buzzing to

French troops with Captain Gerard's folding bicycle

warn of their coming. He was right, but 45 years ahead of his time. Folding bicycles, used by German and British paratroopers, would play their most significant role in World War II.[41]

Topographers and Telegraphers

In America, Lieutenant Henry H. Whitney found, during marches with his artillery battery, that on his bicycle he was able to move far ahead of the column, make an accurate route map with his compass, explore cross roads, mark out the best campsite for the night, and leave markers for the column commander as to the best roads to take. In particular, his cycle odometer not only recorded distances to within 20 yards accuracy, but was much more efficient than pacing by foot or horse, which required counting steps and noting animals' changing gaits.[42]

He soon adopted the bicycle for assessing road gradients, a necessary task in estimating the rate of movement of wagons and artillery. It was a tedious job, in which the observer frequently had to stop, dismount from his horse, and take sightings along a protractor while noting the angle of a plum-bob on a thread. The Lieutenant had a tinner make him a "bicycle clinometer," consisting of "an arc graduated both ways from zero and swept by a pointer pivoted at the center of graduation, the index being sufficiently heavy to act as a plummet and hang vertically at rest." It was attached to the lamp bracket, and adjusted to read zero when the cycle was on level ground. He could read the clinometer without having to stop, whether pushing the machine uphill or coasting downhill.

Whitney noted that a long hill of gentle mean slope may have a slightly steeper section or two, "where a wagon or artillery piece will stick," but which is not readily appreciated by the eye. With his device, such minor changes in slope could be readily observed and measured. By measuring the length of a slope and its gradient, horizontal distances between contours could be readily reckoned with a table of equivalents. To facilitate the process, Whitney attached a sketch board to his handlebars, with forms for recording notes. Using this setup, he felt that a small corps of map making cyclists could, in a short time, assemble a mass of information about roads, paths, trails and streams that would be of immense value to the War Department. There is no indication, however, as to whether his ideas were taken up by other military engineers.

The Lieutenant's experiments paralleled the immense amount of civilian road map activity under way at the time, around the world. As millions of new cycle owners began traveling away from their local haunts in the 1890s, they sought accurate and detailed information about road conditions, but it was scarce. As a result, the modern road map evolved at this time, along with cycle touring organizations that later became automobile clubs. Australia's George Broadbent, the country's premier road map maker, sold 110,000 copies of his new road maps to cyclists before any significant numbers of motor cars were registered. In America, innumerable regional road maps were produced by individuals, bicycle touring clubs, and commercial interests, and English bicyclists played a crucial role in the formation of the Automobile Association. By the time the motor vehicle hit the road, the way had been well mapped by cyclists.[43]

In August, 1896, *Scientific American* described bicycle-mounted devices that could lay down and pick up telegraph lines. The article included an engraving of a machine invented by Leo Kamm, a German living in London, which was fitted on a cycle's rear wheel. It incorporated a drum, holding about a mile of telegraph wire, which unwound the line as the bicycle rolled

A British Army cyclist telegraph engineer in the Boer War, with a cable-laying reel

along. A bell rang before the wire was completely payed out. The wire passed over a wheel connected to a telegraph receiver; the rider would dismount, fix an earth rod in the ground, and send or receive a message. The wire laying apparatus weighed seven pounds, a reel of wire 10 pounds. The machine carried two extra reels under the crossbar. It was demonstrated at Aldershot and other military gatherings.

In San Antonio, Texas, Captain R. E. Thompson, of the U.S. Army Signal Corps, invented and tested a similar device. It would not only lay down line, but the rider could "reverse the action" and pick it up as well. On a trial recovery run the wire was spooled evenly with constant tension, even though the rider often deviated several feet to either side of the line lying on the ground. The reel held a third of a mile of cable, which was reportedly unwound and retrieved in a total of two minutes. There was, unfortunately, no illustration. The San Antonio trial must have been carried out in late 1895, as an Australian cycle journal reported it in February 1896. By then signal units in Arizona and Texas were using bicycles for patrol and repair work along telegraph lines. In an El Paso trial, a lineman mounted his bicycle, pedaled out, completed a repair, and returned, as fast as it "formerly took to hunt up a horse and wagon and get started."[44]

The Cycling War Correspondent[45]

In April, 1897, Wilfred Pollock covered the Graeco-Turkish War as a correspondent for the London *Morning Post*. Later that year he wrote a book, *War and a Wheel*, describing his use of a bicycle to get around. When setting out from London, he had been warned that a cycle would be of no value on Greek roads and did not take one. Upon arrival in Athens he deemed the advice sound after viewing the city's rough streets and crossing his first mountain pass. He concluded that the bicycle "would have meant merely so much dead weight for a mule." However, by the time he reached Larissa he found more good roads than he had expected and had a cycle sent from Athens. Over the next few weeks Pollock found bicycling to be the fastest means of getting around Greece. He rode and pushed it for 200 miles over "severe country" and carried it with him on trains and boats.

The country roads generally proved to be much better to pedal over than the cobblestone streets of the towns, and among the smoothest surfaces were the sheep paths alongside the roads. Pollock regularly saved hours of travel time compared to his equestrian colleagues. He concluded that a 40 mile cycle journey, in which he pedaled the good stretches and pushed over the rough parts, was much less exhausting than riding a horse, and much faster. Twice the bicycle enabled him to be the only journalist to make it to distant telegraph points in time to file battle reports for his paper's next day edition.

The ability to throw the machine on trains and, especially, small boats that could not take horses, gave him maximum flexibility in getting about the country and along the coast, with a minimum of fuss and arrangements. During two panicked retreats by the Greek Army and civilians—at Velestino and Turnavos-Larissa—other correspondents had to abandon gear that usually accompanied them by wagon. Pollock, who had only what he could carry on his bike, simply hopped on and took off. He expressed little sympathy for his colleagues during the routs: "every correspondent of sense who had to accompany the Greek army must have known pretty well before he started that he would be on the losing side."

The machine itself gave him no trouble. However, the Athens cycle shop had fitted only one wheel with a heavy duty tire. The other was a thin racing tire, which suffered numerous gashes from sharp marble stones. Eventually Pollock replaced it with a used heavy duty tandem tire. He also broke one of the two saddle springs while pedaling across a rough, partially plowed field, was unable to repair or replace it, and found subsequent riding sometimes "horribly uncomfortable." He left the bicycle wherever he wished and did not worry about theft, although he did have his tool bag stolen three times. He found that the cycle, rarely yet seen in most of the areas where he rode, tended to cause mules to shy. Consequently, he usually dismounted or slowed when approaching them. The situation was undoubtedly not helped by two sleigh bells which, inexplicably, he carried on the cycle.

Battlefield Tactics

It was widely felt that on the battlefield cyclists would be most valuable for "exploring, reconnoitering, harassing, and expeditionary patrols, all demanding celerity and silence of movement." Field results confirmed that assessment. In Austrian maneuvers, in 1897, a unit of 24 cyclists eluded enemy sentries by sneaking themselves and their machines through meadows and woods, and then firing upon unsuspecting artillery units. When horsemen attempted to ride up a hill after them, the cyclists quickly pedaled out of range down the other side. That same day, while protecting dragoons, the hidden, dismounted cyclists likewise surprised the opposition. The overall assessment was that on both occasions "their sudden and unexpected appearance struck confusion into the opponent."[46]

Numerous soldiers practiced getting their machines and other equipment over high walls and other obstacles. The Connecticut National guard experimented extensively with fording streams, carrying wounded on the machines, and riding across pastures, fields, broken country, and through woods. They suggested that a bicycle could be effectively hidden by throwing it into a stream or lake, and if in danger of falling into enemy hands, could quickly be rendered useless by removing the chain, tearing out spokes, flattening tires, or bending frame components.[47]

The speed and flexibility of cyclists vis a vis the cavalry were demonstrated facts, but there was always that nagging question: How would cyclists defend themselves against a battlefield cavalry charge, if necessary? The answer was the *zariba* (or *zereba*), a term reportedly derived from an Arabic word to describe a thorn enclosure. In 1889 the *Illustrated London News* described the tactic. A group of cyclists would dismount, up-end their cycles, and form a square or circle, with the machines slightly overlapping one another. The soldiers would lie or kneel inside, balancing their rifles upon the frames to steady their aim. As the horses approached, the cyclists would set their wheels spinning, and the reflected sunlight would ostensibly frighten off the animals. The resultant cycle fence formed "an obstacle which few horses, if any, would face." The soldiers, "in perfect security," could pick off the advancing horsemen "with deadly effect." The cyclists' position, "so far as mounted horsemen are concerned, is practically impregnable." Even if cavalrymen did manage to get their animals to charge, they would be tripped up by the wheels and frames. As an American soldier noted, "from personal experience I can assert that there is nothing that will tie up a horse's feet so effectually as the steel spokes of a bicycle." However,

A "zariba" formation to defend cyclists against charging horses

if the photographs of zaribas are an accurate indication of their dimensions, any charging horse that might fall into one would land directly upon the enclosed cyclists.[48]

And with so many potentially dangerous cyclists pedaling to or about the battlefield, the Kaiser's army trained Alsatian dogs to chase down enemy cyclists and pull them off their machines.[49]

Military Attitudes and Rates of Adoption

Many observers in the 1890s—both military and civilian—felt that far more soldiers should have been using bicycles than were, and that more trials should have been carried out. Their lobbying was persistent, sometimes intense, and often insensitive. The editor of *The Australian Cyclist*, in pressuring the colony of Victoria to adopt cycles for the militia, cited recent favorable reactions in England and asked "how is it that Victorian authorities lag so lamentably? Our present Commandant, Sir Charles Holled Smith, is a smart soldier, and would, we feel sure, echo the opinion of the head of the British Army." Sir Charles did not.[50]

In America, bicycles were in daily use in a number of military installations, although various accounts at the time suggest that the soldiers had to furnish the machines themselves. In contrast, at least four state National Guard units and various militia districts formally adopted them. Contemporary accounts suggest that there were probably less than 1,000 official military cyclists in total in America at mid-decade. Yet, one officer judged that the interest taken in bicycles by the enlisted men at his post was so great that, if offered the opportunity to buy one at cost and use it on duty, "fully 10 percent [of soldiers] would be mounted within a month." All it needed was authorization, and would cost the army nothing. The failure of officials to act led him to comment that, "while being the most progressive nation on earth in matters civil, we are among the most conservative in affairs military. Old ideas are regarded almost as a fetish; we shrink from making new experiments." His view was echoed by another officer, who attacked the lethargy of the "cold military conservatism, which judges too frequently almost with ferocity and without investigation."[51]

In England, a number of influential military officers supported the use of the bicycle. General Sir Evelyn Wood recommended in Parliament that the country should encourage the

A cycle ambulance crew from the London Cyclist Battalion

raising of 20 volunteer cyclists. He noted that "during the whole time I was in India during the Mutiny, I do not remember, except when actually in the hills for three or four days fighting, one day's march, or any one fight in which we took part, where cyclists could not have been used with the greatest possible advantage." One English journal felt that the Duke of Connaught and Lord Roberts, along with General Wood, had "contracted a bad attack of cycling fever."[52]

By 1898 British Volunteer regiments totaled 3,488 cyclists, who had to supply their own machines. The 180 cyclist sections included 10 in Wales, 34 in Scotland, 39 in English metropolitan areas, and 97 in the provinces. The largest was the 26th Middlesex (Cyclist), with 116 members. However, there were no bicycle mounted soldiers in the *regular* British army that year, although the government did order 100 machines "for experimental purposes." A noted cycling writer of the era acidly observed that finally "the War Office—and it takes a long time for the War Office to find out anything—has discovered the merits of the new [safety bicycle] invention", by then a decade old. The War Office specified that bidders had to use parts from a particular Birmingham cycle manufacturer. Competitors complained that it was unreasonable for companies such as Sunbeam or Singer to have to build their machines with a rival's components. Eventually the matter was raised in Parliament, and manufacturers were assured that it would not happen again. The bicycle had been drawn into the British military-industrial complex.[53]

It is possible to provide only a crude estimate as to how many military cyclists served in various countries during the 1890s. The most comprehensive survey was prepared by Lieutenant Henry H. Whitney, in the United States, in 1895. Combining it with other available information, it would appear that, by the end of the century, there were perhaps 20,000 bicyclists serving in some military capacity worldwide. That would include British Volunteer regiments, U. S. National Guard units, and equivalents elsewhere. The majority appear to have been involved in messenger or related work.[54]

In France the bicycle was well entrenched, its regular army the most committed military user in the world. It was reportedly introduced into the forces by ministerial decree in July 1887. By 1895 the headquarters staff of each army corps had 19 bicyclists, each infantry division 11, and each cavalry division eight. As well, a small number of machines was allocated to the staff of all infantry and cavalry brigades and regiments, each ambulance corps and division headquarters, every engineer company, and each field bakery and battalion of chasseurs.

Lieutenant Whitney's report contained some surprises. For example, in Germany only 102 machines were officially in use, although the 1894-1895 army budget allowed for 728 more. Whitney attributed the small number to the fear of some cavalry officers that the machines might displace their messenger and scouting functions, and they consequently used their influence with the Emperor, who "is known to be very partial to the cavalry." Whitney thought that mountainous Switzerland would have been "one of the last countries" to adopt the bicycle. However, to his astonishment, a bicycling school that opened in 1892 had trained 198 officers and men in riding techniques that year alone. By 1895 the Swiss army headquarters staff had 15 cyclists; each of the four army corps had five; each division 15; and each brigade and district had a cycling corps.

Elsewhere in Europe, at mid-decade, Holland's army had 75 places "opened to cyclists," who received an allowance for providing their own machines. Spain's initial 19-man cyclist unit of 1890 had been expanded to include "cyclist detachments" for engineer and infantry regiments, "as soon as they are able to provide

A military velocipede invented in America in 1899. It was steered by the knees, freeing the hands for shooting, geared at 3:1, and fitted with an equipment rack.

for the expense." In Belgium all military cyclists were trained at a regimental bicycle school of the Carabineers, opened in 1890. And in Italy, whose army was traditionally credited as being the first to adopt the bicycle, in 1870, there were four bicycles assigned to each regiment of infantry, grenadiers, sappers and miners, engineers, and cavalry. He provided no figures on the number of machines in use in Austria or Russia, although in the latter case bicycles had been incorporated into the army in 1891; two were allocated initially to each field infantry and rifle regiment, to be increased to four when finances permitted and the arsenal was able to manufacture them.[55]

A military cycle on the Trans-Siberian railway line, details and date unknown

A factor that almost certainly worked against the bicycle's military adoption in the 1890s—but one that is hard to assess—is that bicycle soldiers in parades look relatively sloppy in comparison to other soldiers. Cyclists riding at a parade march rate have to turn their wheels back and forth slightly to keep their balance, and can maintain neither straight ranks nor parallel lines. Also, the normal riding position leaves one slightly hunched forward. Some early films of cyclist soldiers on parade clearly illustrate that they simply do not possess the sharp, snappy look of marching infantry. A common response was to have cyclist soldiers push their machines along or march without them.

By late 1899, the basic characteristics of the bicycle and the capabilities of the bicyclist had been reasonably well demonstrated. How they might fare in war, however, was another matter. Those searching for that answer faced a dilemma common to many: since the Franco-Prussian war, a quarter of a century before, there had been no major testing ground for studying the wartime effects of modern weapons and supporting technology. And that intervening time had seen developments which had the potential to radically alter the nature of warfare. The bicycle, the forerunner of a mechanized and motorized personal transport revolution that was about to sweep the world over the next two decades, was but one of many unknown elements.

Major John M. Macartney voiced the view of many at the time, that bicyclists and cycle technology would have little effect upon future warfare: "Mounted on an ordinary bicycle, a soldier is powerless for offence or defence, and falls an easy prey to a man on foot. For this reason cyclists have never been seriously regarded as offensive factors, notwithstanding their own assertions to the contrary." Even those who gave them some credence often believed that "cyclists only differ from infantry in having cycles." While those perspectives ignored the uniqueness and flexibility of "bicycling", they paralleled another viewpoint at the time: that a machine gun only differed from a soldier in that it fired faster. The military at the end of the nineteenth century tended to assess matters essentially upon the basis of how previous wars had been fought. But as history has shown—and would soon reiterate in South Africa—that is rarely the way the next one happens.[56]

2. The Boer War

Among the questions likely to be settled by the present war is that of the use of cyclists in the field.
        COMMANDER C.N. ROBINSON, *THE TRANSVAAL WAR ALBUM*, 1902

It was in South Africa that the bicycle was first wheeled into the military spotlight. There, several thousand British soldiers, Boer commandos, messengers, and scouts pushed, pedaled and carried their machines amongst the rugged hills and mountains of Natal, across the high veldt and kopjes of the Transvaal and Orange Free State, and through the sand and over the salt pans of the eastern Kalahari desert, in Bechuanaland. As well, bicycling technology was adapted to laying telecommunications lines, carrying the wounded, and patrolling railway lines with specialized war cycles. When it was all over, many military thinkers would take a fresh look at the machine's capabilities and potential.

The Boer War proved to be Britain's longest, costliest, bloodiest, and most humiliating between 1815 and 1914. The "last great imperial war", as Kipling described it, taught the British "no end of a lesson." The rest of the world also learned some lessons, for the Boer War saw a number of technological elements come together: observation balloons, armored trains, telephones, the telegraph, smokeless powder, machine guns—and bicycles. Each had already been the subject of varying degrees of experimentation, development, practical use, and prophecy, but they were first collectively trialed in the South African military crucible.[1]

Some inventions, such as smokeless powder, had immediate and far-reaching effects upon battlefield tactics, while others, such as the machine gun, never did find their niche in South Africa—either in practice or in the military mind of the day. Yet other devices, such as the telegraph, significantly influenced matters: the War Office in London knew the results of some battles within minutes, and officers were occasionally second-guessed by

Cycle technology adapted for stretcher use

newspaper readers even before they had written an official account for their superiors. It was a new kind of war, as each tends to be.

Jameson's Wielrijders

The Boers (mostly Dutch Calvinists) settled the Cape Province in 1652. One hundred and fifty years later the British took possession of it because of Cape Town's importance as a naval base. Their subsequent relations with the Boers were uneasy. In 1835-37 the Boers marched inland to establish their own colonies, the Orange Free State and Transvaal. Following the discovery of diamonds and gold on Boer lands in the late 1800s, pressure was put upon the British government to annex those colonies as well. By late December, 1896, tensions were high, and powerful mining interests attempted to initiate a rebellion against the Boer government, led by Dr. Leander S. Jameson.

Jameson's base was in Bechuanaland, near the Cape Colony and Transvaal border. His force consisted of some 600 men, mostly Rhodesian police belonging to Cecil Rhodes's Chartered Company, which administered Rhodesia for the British government. His objective was to march to Johannesburg and, with the assistance of British residents and other foreigners, rise up against the Boer government. The raid was a fiasco, but Jameson's march across the Transvaal provided a taste of the nature of warfare that would come in another three years. The mounted troops, riding almost nonstop across the veldt, took four days to cover 170 miles. They were harassed by the Boers who were virtually

A cyclist messenger from Jameson's Raid, showing the saddle pillar in which he hid dispatches, 1896

invisible in their brown work clothes, rose up out of the ground like ants, fired, and then disappeared. The field guns and Maxims of Jameson were of little avail in such circumstances. It was the forerunner of the guerrilla warfare principles that would be finely honed by the Boers' Christiaan De Wet in 1900.[2]

While Jameson's horse column trod slowly across the landscape, express cycle riders, in contrast, "scoured the country at will" at the rate of 15 miles an hour, carrying messages between his column and Johannesburg. A noted racer, J. D. Celliers, was one of a pair of Johannesburg cyclists who delivered a message to Jameson. On their return ride, carrying his reply, they were intercepted by Boers and abandoned their machines. The despatch was recovered some months later, still hidden in the frame of the bicycle. Another cyclist, E. M. Rowlands, also carried a message to Jameson, concealed in the saddle pillar. The machine's manufacturer, Osmond Cycles, ran worldwide advertisements showing him pointing to the hidden missive. The exploits of the express cycle riders were reported at great length, within days, in various local and overseas newspapers, magazines and cycle journals. The net effect was to widely publicize the speed of the machine and its military potential on the South African veldt.[3]

During the last years of the 1890s the bicycle became thoroughly established in South Africa. The machines were profitable items to carry on ships, and thousands of American and European models found their way to the Cape. By 1897 South Africa was second only to Australia as an export destination for English bicycles, at which time there were an estimated 8,000 to 9,000 thousand in the Johannesburg area alone. Speed limits were legislated, cyclists were prohibited from riding on footpaths, and the Town Council threatened prosecutions. Similar steps were taken in Kimberley. Riders accumulated considerable knowledge about the cycle's capabilities in the South African rural environment, and when the Boer War finally erupted in October 1899, there was a core of experienced cyclists ready to serve on both sides.[4]

The Twentieth-Century War

At the outset of the war, the Boers' leadership and fighting abilities were grossly underestimated. Joseph Chamberlain, the British Colonial Secretary, thought it would be over quickly. He was

This dispatch rider had to carry his and his horse's food supply for the journey from Bloemfontein to Kimberley

wrong. The Boer forces swept into the mountainous British colony of Natal on October 11, 1899, from the Transvaal and the Orange Free State, and over the next six weeks dealt General Buller a series of stunning reverses. They quickly drove deep into southern Natal, laying siege to Ladysmith, while their comrades surrounded the isolated mining towns of Mafeking and Kimberley, several hundred miles west. The British were suddenly confronted with a war over a vast area.[5]

Worse, they faced unorthodox tactics. At the battle of Magersfontein, on November 28, the British Army's traditional artillery barrage was followed by the cavalry and infantrymen charging upright through barbed wire entanglements. In contrast, the Boers lay in trenches and hid behind rocks, from where they directed deadly accurate rifle fire. Thirteen days later, some five miles north, the point was reiterated. The war definitely would not be over by Christmas, as many had suggested.

Nor would it end soon afterward. On January 24-25, at Spion Kop, 10 miles west of Ladysmith, the Boers and British clashed in what turned out to be "an acre of massacre," with bodies three deep in the shallow trenches. General Buller's 19th century army was "learning how to fight a twentieth-century war." The General himself never did, and was replaced by Lord Roberts, with Lord Kitchener as his Chief of Staff. The British effort was then radically increased, and eventually some 450,000 British and colonial soldiers served in the war, against only 87,000 Boers. Under Lord Roberts, the British achieved their first major victory, in late February, at Paardeberg, assisted by a balloon sent up 1,000 feet. From its basket an engineer drew detailed sketches of the Boer positions, dropped them to the ground, and they were used to direct artillery fire upon the Boers. It was truly a new kind of war.[6]

By mid-year the British had regrouped and lifted the sieges of Kimberley, Ladysmith and Mafeking, and had captured Bloemfontein and the Boer capital, Pretoria. The Boer president, Paul Kruger, fled to Europe and the end of the war appeared in sight. The Boer commanders, however, refused to surrender. At the same time, their recent large losses during field battles against the numerically superior British convinced them that they would have to switch to a campaign of guerrilla warfare. The initial British attempts at finding and riding down the highly mobile guerrillas in their native terrain proved frustrating. Subsequently, the British turned to burning farms, raiding stock, and gathering the non-combatant population into concentration camps in an effort to eliminate the Boer guerrillas's food supplies, intelligence sources, and civilian support.

By early 1901 the British developed a successful strategic policy, based upon the fortified blockhouses that they had built to protect the railway lines from Boer raids. The blockhouses were the brainchild of a major in the Royal Engineers, who hit upon the idea of adapting the basic cylindrical corrugated iron water tank, used throughout South Africa. He put one cylinder inside another of slightly larger diameter, filled the space with rock, cut small portholes, and added a conical roof. It was cheap and effective. Although it offered protection only against small arms fire, the Boers had little else from 1901.

Each blockhouse, staffed by a handful of men, was within sight of the next, usually from half to three-quarters of a mile apart, with overlapping fields of fire. Multiple strands of barbed

A blockhouse on the Delagoa Bay railway line, manned by Argyll and Sutherland Highlanders

wire were strung along the railway lines between the blockhouses, together with trip wire signals and trenches to prevent wagons crossing should the barbed wire be cut (the early wire was gradually replaced with a thicker, higher tensile version). The blockhouses were linked to one another by telephone and telegraph, and the railway lines patrolled by armored trains. The latter were equipped with guns, searchlights and a team of linemen and telegraphers who, when the train stopped, could quickly connect to the telegraph system. The result was a network of blockhouse-and-wire barriers crisscrossing the countryside, which was extremely difficult for mounted Boers to traverse.

An armoured train truck with a rotating gun turret. The two-man railway cycle and bicycle were for scouts and line patrols

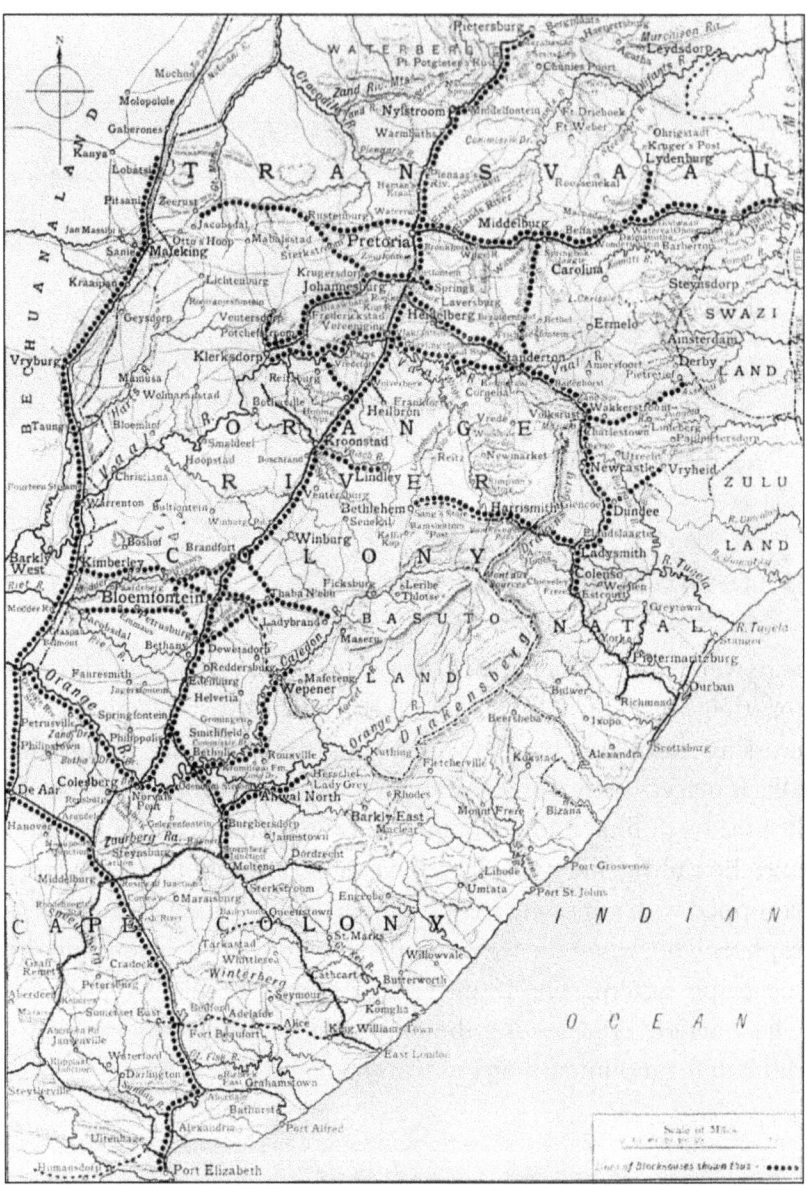

A map of the partially completed blockhouse network that eventually partitioned much of South Africa

Ultimately the network was extended to encompass roads and the open veldt. By the end of the war there were over eight thousand blockhouses interspersed along 3,700 miles of corridors. The original railway defensive network was thus effectively converted to an offensive system as the British forces undertook systematic "drives" across each of the countryside's wired off compartments. The Boer leader, de Wet, disparaged it, but it gradually took away the great Boer ally, open space. It took two years for the British to run the remnant commandos to ground. Eventually "there came a day when there was nowhere left to go, and no Boers free to go there."[7]

The Railway War Cycles

The most interesting use of bicycle technology in the Boer War was the creation of "war cycles" to patrol railway lines. The devices were first constructed by Donald Menzies, at his Bloemhof Street workshops in Cape Town, soon after the start of the war, in response to a request from the general manager of the Cape Colony's railways.[8]

Menzies started with a pair of four-man bicycles, which he aligned side by side with a metal framework. Each cycle wheel

The two-man and nine-man railway cycles at the Menzies factory in Cape Town

was then fitted with a pair of flanges, turned from one-eighth inch steel plate. The inside flange was larger than the cycle wheel and was used to keep the war cycle on the track, as in traditional railway bogeys. The outer flange, the same diameter as the cycle wheel, was for strengthening the bicycle rim. Wooden blocks were used to bolt the two flanges together. Given the strain that would be placed on the wheels, and the extremely high temperatures the rails could reach in the hot South African climate, Menzies decided to wire solid rubber tires on to the rims, rather than use pneumatic ones.

The eight-man machine, weighing 250 pounds unladen, with a gear ratio of eighty-four and a ten foot wheel base, was finally taken out for its first shakedown—and there was much to shake. The loaded weight was about 1,500 pounds, and it was fixed wheel with no separate braking system. One result was a high risk of broken chains if either of the two sets of riders tried to start too suddenly. Conversely, the kinetic energy built up in the heavy machine at speed posed the same problem when trying to brake too quickly by backpedaling. The front forks had to have extra stays fitted to keep them rigidly in line, because when rolling over the points at which rail lines crossed each other, the wobbly front wheels tended to derail. The solid rubber tires made pedaling difficult. The vehicle reached 20 miles per hour only with great effort over a short distance, and had an effective cruising speed of only 10 miles per hour. Steel wheels were tried but proved to be only marginally faster, and their noise "rendered them impossible." In summary, the eight-man war cycle was not practical unless it could be fitted with pneumatic tires, and the risk of them bursting was considered too high.

A two-man railway cycle. Some 20 were reportedly built for patrol work

Menzies took the lessons back to his factory. He concluded that a successful machine would have to have pneumatic tires, and be light and completely rigid. He opted for a two-man version as a viable alternative. Rather than modifying bicycles, he built a specialized frame from cycle tubing. Each corner had a rigid fork system for the wheel, and the whole structure was diagonally braced. Every joint was brazed, with no nuts or rivets used. He debated using an axle on the rear driving wheels, but stayed with a traditional chain drive and a gearing ratio of 80. The wheels had lighter flanges, made of one-sixteenth inch steel, fitted more than an inch from the pneumatic tires. This allowed the tires to maintain full contact with the rails, yet negotiate crossing points and curves with safety.

The two-man version weighed 60 pounds unladen, but Menzies felt that it could safely have been made considerably lighter. He and his colleague, J. G. Rose, one of Cape Town's top racing cyclists, tested it. It proved faster and smoother than an ordinary roadster bicycle and, observed by a correspondent from *The Lancet*, achieved a speed of 30 miles per hour for a short distance. Also, at speed, it safely traversed all the points and crossings in the Cape Town railway yards. It coasted downhill quickly, and was relatively easy to pedal up the same grade.

The war cycle ran "all but noiselessly," and could carry two passengers and a Maxim gun. Painted khaki, it was "difficult to see even at comparatively short distances," and was light enough to be easily lifted from the line. Because it did not need steering, the riders were free to shoot and observe from the smooth riding platform. With its fixed wheel, it could also be pedaled forward or backward. In March 1900, a photograph was published showing the eight-man and two-man prototypes in Menzies's workshop.[9]

A writer in 1901 mentioned that a score of the two-man war cycles had been constructed by then, although there is no known report on their performance. One photograph of an armored train shows both a two-man war cycle and a bicycle stacked atop a gun carriage, with the caption indicating that they were "for scouts." In July 1902, a Cape Town newspaper printed photographs of five war cycles used for routine patrol along the Pretoria-Pietersburg railway line.[10]

Pictures of the war cycles (especially the eight-man version) have appeared in several military history books over the years, usually with insubstantial and sometimes inaccurate captions. Consequently, there are three bits of conventional wisdom that have grown up around the machines which need to be corrected.

D. R. Maree stated that the war cycle was introduced into South Africa by the Royal Australian Cycle Corps, a "prototype" of which he indicated is on display at the Fort Klapperkop Museum. However, no Australian War Memorial archival materials refer to any "Royal Australian Cycle Corps" or war cycles. All extant evidence points to the first war cycles having been built by Donald Menzies, in Cape Town. Correspondence with the South African National Museum of Military History indicates that the "prototype" at Fort Klapperkop is actually a reproduction made for a film.[11]

Cape railway yard with cycles

In the first edition of his history of the Boer War, *Good-bye Dolly Gray*, Rayne Kruger stated that the war cycles represented "the first time pneumatic tires ever took to the railways." That is not the case. In the 1890s several devices were available that enabled a bicycle to be ridden on train tracks, including some versions that even had seats for passengers. Although most were generally unsatisfactory in use, the American retail company of Sears, Roebuck offered various of them for sale through its mail order catalogues.[12]

Occasional writers have left the impression that the war cycles could routinely travel at 30 miles per hour. As the contemporary accounts noted, that speed was only reached for a very short time on the two-man cycle, and was not a cruising speed. Twenty miles per hour was the maximum speed reached on Donald Menzies's eight-man version, and that for only a few moments.

Boers and the Bicycle

The most famous Boer advocate and user of the bicycle was Danie Theron, a young Krugersdorp lawyer-turned-guerrilla fighter who was instrumental in creating the 108-man Wielrijders Rapportgangers Corps (the wheelriders rapid movement corps). It was established in mid-September, 1899, as war loomed on the horizon. Seven Wielrijders sections, ranging from eight to 18 men, were dispersed throughout the Transvaal and Orange Free State to provide communication links between Boer groups.

Boer Commandos and their bicycles

Theron, born in Kroonstad of Scottish parents, was "boyish in appearance, of slight build," and spoke English without a trace of Afrikaner accent. That combination enabled the English speaking "lad" to get through British lines with considerable ease. It was post-war legend amongst Free State Boers that a large number of British troops captured in a trap in early June, 1900 were led into it by Theron, riding a bicycle. Lord Roberts considered him a major thorn in the side of the British until he was killed near Krugersdorp, three months later.[13]

Operating in their homelands, and provisioned by locals, the Boers tended to stay out for days at a time when on patrol, occasionally sending someone back to report. Since the long-range rifle made invisibility a virtue, the Boer cyclist despatch riders—who were fast, quiet and raised little dust—soon proved their value not only for communications, but for scouting and patrol

The Boers' Lichtenburg commando, with bicycles, which captured Vryburg

work. It was suggested by Sir Frederick Maurice, the British military historian, that the Boer cycle despatch riders were "disliked" by their horse-mounted comrades. Others, though, commented that while many burghers initially looked down upon the cyclists, once they witnessed the performances of the despatch riders and scouts, they developed respect for them. Eventually several Boer fighting groups besides Theron's Scouts (as they became known), used bicycles.[14]

Ultimately, the Boers relied upon bicycle riders in many circumstances. For example, during their retreat from the Orange River in March 1900, they drove 750 ox wagons over 140 miles. The convoy stretched along 15 miles of road as it traveled north along the Basuto border, via Wepener. While Boer horsemen protected the slow moving convoy's flanks, cyclists pedaled up and down the lengthy column maintaining communications. Although enough has been written by and about Boer cycle use to gain some indication of the nature and extent of it, there is nothing on the scale of British records, reports, and diaries with which to retrospectively assess the situation in any detail.[15]

The British Experience

The British quickly adopted bicycles and cycle technology for many purposes, ranging from pedaling paymasters to stretcher cycles. Indeed, the use of bicycles was anticipated. Several months before the outbreak of war, a secret analysis of road conditions in the Cape colony requested an assessment of the rideabilty of macadamized and veldt roads for cyclists. A typical comment noted

A detachment of cyclist orderlies at Cape Town military headquarters

that "the road from Cookhouse to Somerset is a fair bicycle road, the stationmaster can travel it without dismounting." As with the Boers, the most significant British use of the bicycle was for communication. From the outset of hostilities, many war telegrams had the note "per cycle despatch rider" added to the document, to expedite local delivery. Racing cyclists who joined the British forces to serve as dispatch riders and scouts were widely praised for their work, and at the sieges of Kimberley and Mafeking orderlies used bicycles to carry messages and orders.[16]

The best description of bicycle usage by British troops was provided by J. Barclay Lloyd, a cyclist with the 1,000-strong City Imperial Volunteers. They saw action across both the Orange Free State and the Transvaal. Prior to embarking in England, the C.I.V. cyclists replaced the standard military valise with a capacious rucksack, recommended by an alpine climber. Each rider carried on his person a 100-round bandolier, full haversack and water bottle, a belt with a pouch and bayonet, a carbon filter attached to a lanyard, a pair of field glasses, a knife and a compass. On the bicycle was a mackintosh rolled up and tied to the handlebars, a rifle in a bucket attachment, and on the back a carrier that held a rolled great coat, blanket, waterproof sheet, and the rucksack, inside which were clothes and a mess tin. The alpine rucksacks could be worn by the rider or strapped on the cycle, as circumstances required. By loosing a single strap, they could drop the heavy kit from the back of the machine, "ready to go on light quick work."[17]

Immediately upon disembarkation in South Africa, many C.I.V. cyclists were assigned to orderly work between the docks and camp, their prime function throughout the war. When the 1,000-man force was on the move, bicyclists served as despatch riders up and down the column, and between adjacent communities and military installations. When the cyclist battalion was rostered at the rear of the column, the men would "ride a tortoise race behind the regiment in among the water-cart mules." When heading the column, they would pedal out to the cavalry screen, assist in scouting suitable roads, then pedal back with directions.

On sandy soil Lloyd found that mounting and getting the laden machine underway was sometimes difficult. Nevertheless, over a wide variety of terrain and across diverse riding surfaces he could still far outpace infantry troops. The cyclists were better off in that they could carry on their machines "without hindrance or discomfort, far more than a man can bear upon his back." On one 90-mile infantry march to Bloemfontein, which took seven days, the accompanying cyclists pedaled leisurely along. As infantrymen began falling out from illness and injury, C.I.V. cyclists were ordered to pedal up and down the two-mile long column and assist their marching comrades to reach the adjacent railway line, where passing trains picked them up and conveyed them on to Bloemfontein.

On the field of battle, when a Boer position was charged, there was usually "a cyclist or two with the leading rank, while others will be scattered over the field conveying messages by Kaffir tracks [native footpaths] or across the veldt." On one occasion, when his regiment was engaged in a running fight with Boers, Lloyd had alternately to carry and drag his heavily laden machine up and over boulders to the top of a steep kopje; it was then that he concluded that "a bicycle is not an unmixed blessing."

A group of cyclists from the City Imperial Volunteers

When their camp lay within 30 miles of a large town, the cyclists had their busiest times. They would convey mail, telegrams, dispatches, and returns of killed or wounded to the major center, and pedal back with stores, groceries and money for the payment of troops. Although such round trip journeys could easily be done in a day, they sometimes rode at night to minimize encounters with enemy patrols.

One matter Lloyd complained about was the tendency of officers to miscalculate distances, even when they had detailed

Bicyclists repairing their machines while in camp

maps available. He never could figure out why they did so. Distances measured by him and his colleagues with their cyclometers routinely disagreed with the official estimates, often by half. On occasion it caused great stress for the infantrymen and once resulted in the unnecessary death of several animals when the officers tried to force the column to cover too much ground over a given period.

Horses suffered a high mortality rate in the war. During the blockhouse period, when fewer animals were used, many, like these at the remount depot near Bloemfontein, had to be destroyed

Lloyd found that much of the veldt was rideable, but was covered by a thinly growing thorny scrub which tended to puncture all but the stoutest tires (the long, sharp mimosa thorns came in for special mention). He never noted whether any of the riders tried the Australian-developed, thick-cased Dunlop Thorn Proof tires which were highly resistant to puncture. Those tires had in fact been developed in Australia partly in response to the incidental importation into that country of a native South African thorny plant. He found travel along railway lines relatively pleasant by pedaling on the adjacent paths used by maintenance crews. However, on occasion he rode between the rails when the ballast stones were not too large and covered the sleepers sufficiently for a smooth ride.[18]

The roads and tracks sometimes were very bumpy, and with many sandy patches which could be particularly difficult to pedal through. Overall, however, Lloyd concluded that "the bicycle is a most useful method of progression," with the large number of "fine roads," good veldt tracks and Kaffir foot paths making the widespread use of the machine feasible and successful. A witness at the Royal Commission on the War in South Africa testified that he was "astonished" by the way the City Imperial Volunteers cyclists got about on the veldt. He was equally amazed by the fact that his servant pedalled messages through the very hilly Natal countryside to Pietermaritzburg at the rate of 50 to 60 miles a day, in the midst of the rainy season.[19]

The Colonial Cyclists

Aside from British troops, many colonial volunteers served in South Africa. As Lord Roberts commented, "they could find their way about the country better than the British cavalryman." In that respect the Australian horsemen were second to none: "no other troops had lived under conditions so similar to those of the Boers, or knew better a country of such wide distances."[20]

The same applied to cyclists. The West Australians, in particular, had more experience than anyone else in the world at that time in using the bicycle over vast distances in remote, arid areas. The "special" express riders routinely rode 100 miles a day when delivering mail over the several-thousand-mile bush track network that served the colony's desert mining communities. Not surprisingly, the West Australians had strong views about the potential effectiveness of their bush cyclists in South Africa:[21]

> There can not be any question as to the value cyclists from this colony would be in scouting and despatch work. One cannot help thinking that given the opportunity, some of our "special" riders and overlanders would perform work which would come as a revelation to even those who believe in the utility of the cycle in [despatch] service. From England, wheelmen are being sent out attached to several corps, but to Australians it is somewhat amusing to read that their machines are of the heavy roadster type, and are fitted with mudguards and brakes. A hundred of such cyclists would not be as valuable as half a dozen of the pioneer wheelmen of Australia.[22]

An unknown number of Australian cyclists served in the war. Interestingly, many of them joined the specialist bushmen's units, which officially required every member to be an expert horseman. Bush cyclists were recruited, however, out of respect for their bush lore and proven ability to travel rapidly through the outback. The Queensland Imperial Bushmen's unit, for example,

Soldiers hauling their bicycles across Wagon Drift by means of a rope pulley

included a cyclist company of fifty-one men mounted on Massey-Harris bicycles furnished by the government.[23]

Among the bush cyclists was Arthur Richardson, a West Australian who made the first bicycle crossing of the waterless Australian Nullarbor plain, in 1896, covering nearly 1,800 miles in 31 days. In 1899 - 1900 he pedalled solo around the Australian continent, some 9,000 miles in 243 days, in the longest continuous bicycle ride in the world at that time, through some isolated country and harsh conditions. Richardson joined the West Australian Bushmen's Contingent, which served with Southern Rhodesia's forces, under Lieutenant Colonel Plumer. His troops consisted of "colonial irregulars" from the British South Africa Police, South Rhodesian Volunteers, and the mis-named Rhodesian Regiment, made up mostly of Canadians and Australians. Plumer initially defended Rhodesia against Boer incursions across the Limpopo River, but soon took the war to them by pushing south along the Transvaal-Bechuanaland border.[24]

Plumer's forces included a Cyclist Corps of six officers and 79 men. The first reference to their performance was on the edge of the Kalahari desert, in March, 1900, when Plumer attempted to lift the siege of Mafeking. At one stage he established

Most of Plumer's "Rhodesian Volunteers" were Canadians and Australians

a forward camp at Sefetili, and maintained contact with his main base at Gaborone, 90 miles away, via intermediate cyclist posts at Moshupa, Kanye, and Moshwane. In the flat, arid country, interlaced with sandy surfaces and salt pans, the bicyclists provided rapid communication along the network.[25]

One of his Australian officers, Lieutenant Wynyard Joss, wrote a personal account of the incident in which 20 of his cycle mounted troops fought a group of Boers, killing two and taking two prisoners. He took the opportunity to describe his machine's performance: "We have paralyzed the English officers at the way we ride over the rocks and ruts here," adding wryly, "the only breakages have occurred by the mule teams running over the machines." The circumstances under which the cyclists sometimes operated in the varied South African landscape were indeed amazing. As Joss commented, "if you saw the country we travelled over you would say we were mad to attempt it on bicycles." However, the conditions were in fact similar to those under which many Australian messengers and rural workers had been pedaling for years.[26]

Bicycle Spies and Martial Law

The ability of cyclists to cover great distances quickly and quietly, especially at night, was a significant wartime issue in South Africa, and both British and Boer authorities were forced to take immediate action. The day after the war began, The Standard and Diggers' News announced that permits had to be obtained to cycle outside the town limits of Johannesburg. Despite the fact that "innumerable applications were made, on all sorts of pretexts," none were issued for some time. Eventually all cycles had to be registered, a metal number plate displayed, and lamps mounted at night.[27]

Ten days later the newspaper headlined a story "British Spies on Bicycles." It reported that at the Boer town of Bietfontein, on the Vaal River, "tracks of bicycle wheels have been discovered, and a guard has been appointed with instructions to bring in the riders dead or alive." The Boer concerns were well founded. Prior to the siege of Kimberley, one British intelligence officer, "in the guise of an ordinary cyclist," assessed Boer preparations in the Orange Free State town of Boshof, 34 miles away. The value of the cycle for espionage was so obvious that in another article on October 27, the reporter noted, as a matter of clarification, that "a man named Fraser (not the racing cyclist) has been shot as a spy at Ladysmith."[28]

Eventually martial law was declared by the British. In several communities all bicycles had to be handed in. At Graaf Reinet, Cape Colony, where the loyalty of the large local Boer population was questionable, 500 machines were impounded. In other towns there were either absolute bans on the bicycle's use after dark, or they could only be used with lights. In February 1901, at Worcester,

Bicycles at Graaf Reinet being confiscated by the British from suspected Boer sympathisers

Some of the 500 Boer machines in storage at Graaf Reinet

the local court fined Max Ammer, a German national, ten shillings for riding on a footpath. That same afternoon he was back in court, charged with contravening Martial Law Notice No. 19, which forbade riding between sunset and sunrise without the Commandant's permission. Ammer pleaded not guilty on the grounds both that he did not know the time of sunset, and that he could not speak English, despite evidence that he understood it "fairly well." He was also tried for having talked seditiously, and for making indecent and offensive suggestions concerning how he would treat the British flag. He was found guilty and placed on parole.[29]

One interesting use of British cyclists during the war involved communication by homing pigeons, something that had been tried by French cyclists in the early 1890s. Over 1,000 pigeons were allocated to various troop columns, scouts and loyal farmers to report Boer movements back to 42 homing stations. Some bicyclists operated as mobile scouting platforms. One, named Callister, achieved minor local fame by "cycling 120 miles, gaining a point of vantage, lying hidden for several days, and then releasing birds whenever he saw Boer activity." It could be a risky business. In Boer eyes it crossed that gray boundary between scouting and spying. In 1901 a Lieutenant Wilson and his birds were captured and killed by Boers, as was a scout named Grimes in 1902. Others were luckier. Two men from the Cape Colony Cycle Corps, captured while taking pigeons to Colonel Dornan's column, were released after being stripped of their "bicycles, birds and boots."[30]

Cyclist Numbers

As for the Boers, various sources suggest that bicycles were used by guerrilla units for scouting and messenger work, in addition to the original 108-member Wielrijders Rapportgangers Corps. However, a lack of formal records makes it impossible to estimate how many Boers cycled. In contrast, numerous British Army records and civilian writings refer to many cyclists and cyclist sections that served in the war. These include the 500 strong Cape Colony Cyclist Corps; the 102 men of the "A" Company Cycle Corps of the Kimberley Town Guard; an unknown number of Cape Town Guard cyclists; the 31 riders with the Durban Light Infantry; the 1,000 cycle-mounted troops of the City Imperial Volunteers; the two dozen cyclists of the 1st Royal Dublin Fusiliers; the riders with the Q and U batteries of the Royal Horse Artillery; the D.E.O.V.R. Cycle Corps; the 85 cyclists with Plumer's Rhodesian force; the cycle mounted F Company of the Transvaal Scottish Volunteers; and so on.

Any assessment of how many cyclists served in the war is complicated by two factors. First, while a unit may have been listed as having so many cycle troops, the unit may not have operated at full strength, or not all of the troops may have had bicycles. Second, and more crucially, photographic and written evidence suggest that many soldiers used bicycles under circumstances that would not have led to them being officially classified as "cycle" troops. This applied particularly to those serving as despatch riders, scouts, and orderlies. Major General Sir Frederick Maurice was reported as saying that "three per cent of the active British forces consisted of cyclists." That suggests that upwards of 13,000 soldiers pedaled in one capacity or another during the war. When compared to other sources, collectively, it would seem a reasonable figure.[31]

Questions and Prospects

The performance of cyclists during the early part of the Boer War caused a number of British military planners to take careful note. In January 1901, just after Lord Kitchener assumed command of the

The D.E.O.V.R. Cycle Corps, October 1899

war effort, the 500-strong Cape Colony Cyclist Corps was formed. In March, 1901 Kitchener requested another 1,000 bicyclists from England, and later appealed to the Volunteer Force in the United Kingdom for another eight cyclist companies (a poor response resulted in only two being formed). The intriguing question is, what did Kitchener intend to do with several such bodies of cyclists? Were they only to augment or replace the cycle messengers already operating with various units? If so, what was the logic in creating numerous large cycle corps, whose members would immediately be dispersed to other units? Alternatively, given the guerrilla nature of the war he was dealing with, did Kitchener plan to use the cycle corps in some special way, perhaps to help sweep the countryside as the expanding blockhouse network gradually boxed in the Boers? Unfortunately, there are no known records to clarify the matter.[32]

Whatever the answer, the Boer War saw the bicycle's first significant wartime exposure. In the final analysis the bicyclists and their railway variants acquitted themselves well. The speed and efficiency with which the cycle-mounted messengers and scouts operated in the South African environment vindicated the arguments of military cycling advocates, and influenced thinking about the nature of future warfare. Among those who were previously non-committal about the machine's military potential, a number came to the conclusion that the cyclist definitely had a role to play. In consequence, over the next few years, in England and Europe, there would be an explosion in the number of bicycle mounted soldiers.

3. The War of the Words: 1900-1914

"He saw all these burly, sun-tanned horsemen, disarmed and dismounted and lined up ... their horses unskillfully led away by the cyclists to whom they had surrendered ... these truncated Paladins watching this scandalous sight."
H. G. WELLS *THE LAND IRONCLADS*

On July 30, 1899, *The New York Herald* confined cycling news to the back page, with ship sailings. In 1901, *Scientific American* announced that the cycling craze was over, that the machine had become a commonplace tool. The magazine, which had covered international bicycling developments in detail throughout the 1890s, switched its emphasis to motor vehicles, with only an occasional note on cycling. It was echoing a silence reverberating elsewhere across the land. In newspapers and magazines which once contained not just pages, but whole sections, devoted to bicycles and bicycle racing, cycle news was no longer news. In America, the bicycling boom was bust.

In the United States the issue of military cycling seemed to have been finally resolved, as typified by the Connecticut National Guard. It had been in the vanguard of introducing cycles to signal corps use, in 1891, when Lieutenant Howard A. Giddings talked Albert Pope into providing Columbia bicycles for them. In 1900, in the midst of the collapsing cycle market, Pope recalled all the machines that he had consigned to the corps, and announced that he would not provide any more. The following year Major Giddings handed in his final report as brigade inspector: "There is strong and unanimous desire on the part of the members of the Signal Corps to be mounted on horses instead of bicycles." He recommended the change, and went on to become a vice president of Travelers Insurance.[1]

The bicycle remained far more important outside the United States, but its military use varied widely. In England, Vickers absorbed the lessons from its unsuccessful machine gun tricycle, and in 1899 mounted a single gun on a motorized four-wheeled bicycle carriage. It could carry the gun, rider and 1,000 rounds of ammunition 120 miles, including uphill, and photographs of the self-propelled device appeared in cycling journals around the world. It was a wise move. Members of the London Cyclist Battalion, who had pedaled machine gun carriages many a mile, concluded that the human-powered versions were "perfectly useless for warfare."[2]

In South Africa, following the Boer War, cyclist companies were attached to all infantry regiments as dispatch riders. The first separate cyclist fighting unit in the country, the Transvaal Cycle Corps, was created on October 1 1905, with 31 officers and 415 other ranks. But financial support steadily declined and by 1907 the soldiers were required to furnish their own bicycles. In 1909 it was renamed the Transvaal Cycle and Motor Corps, and in 1913 bicycles were abandoned altogether. The cycle unit's demise was not surprising. Even as the pedal powered railway war cycles were finding their niche during the Boer War, experiments began with motorized counterparts. In 1901, a railway cycle, purpose built to hold a small engine, carried only a single rider.[3]

A motorised railway cycle under trial in South Africa

Thus, to the adage that the bicycle did not need food or water, could be added, "nor gas." In many eyes, however, that was not an advantage, but a sign of obsolescence. For military cycling enthusiasts and the military establishment, it intensified the three basic questions surrounding the bicycle at the start of the century: Was there a future military role for the cyclist at all? If so, what would it be? And how best could it be prepared for? Over the next 14 years the extent of discussion, and the nature of the answers, varied widely from country to country, and led to the involvement of one of the most forward-thinking men of the age.

Australia and the Dunlop Relays

In Australia, the six British colonies became a single country in January 1901. The new government invited Field Marshal Viscount Kitchener of Khartoum down under to advise on a national military structure. In December, 1909, he offered his recommendations. They were based upon the strategic concept that British sea power was the ultimate defense, but that Australia might have to defend itself temporarily. Kitchener proposed a permanent army nucleus supported by a national citizen force. He thought that 80,000 such soldiers and citizen volunteers would be adequate, half to secure larger cities and ports, the rest to operate as a mobile striking force anywhere in Australia. In that context, discussion about the use of cyclists ranged the gamut from forming a full cyclist corps to having none at all. The final decision was made by Colonel William T. Bridges, Chief of Intelligence of the Commonwealth Military Board, and Australian Representative on the Imperial General Staff, London. "If we were raising the forces de novo I should favour the creation of companies of Cyclists," he said, but he felt that it was simpler to add cyclists to infantry companies as dispatch riders. They received an allowance of two pence per mile to encourage training.[4]

In April 1909, in the midst of discussion about bicycles in the Australian military, Dunlop sponsored a major overland relay cycle ride to demonstrate the reliability and value of bicyclists under a variety of weather conditions and over varied terrain. The company hoped to secure profitable tire orders if the military could be convinced to adopt the machine in large numbers. The ride, from Adelaide to Sydney, traversed the Australian Alps

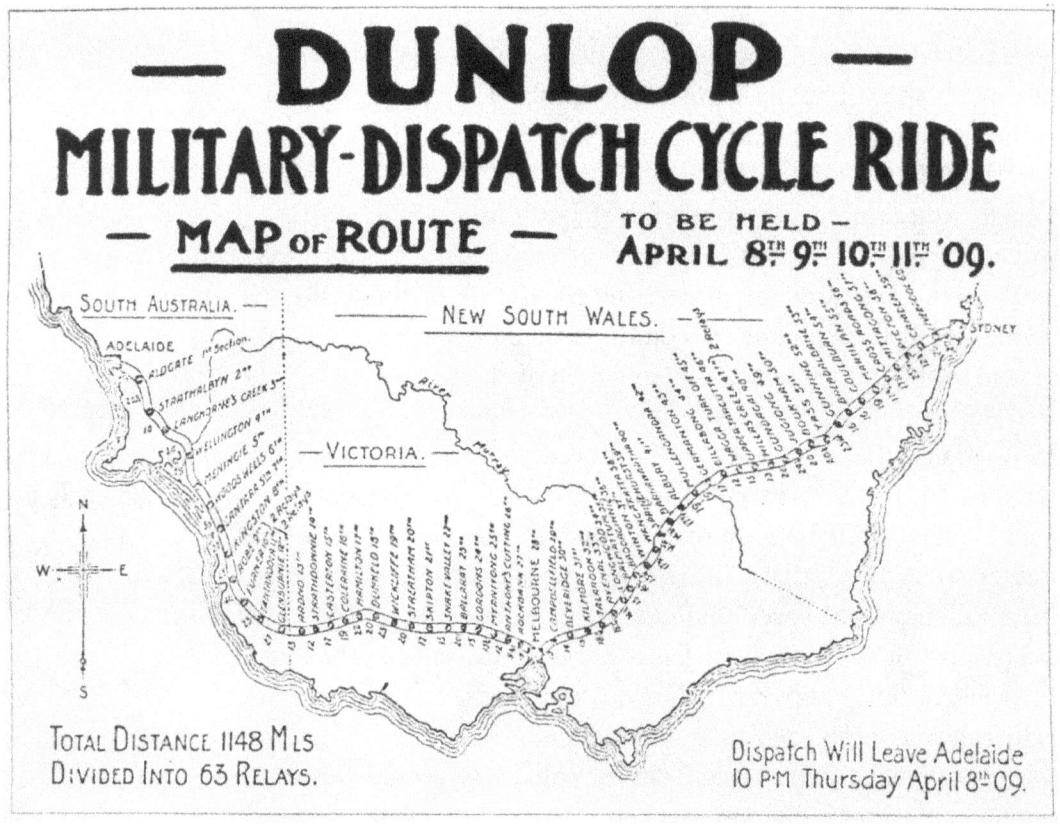

The route of the 1909 and 1912 Dunlop Military Dispatch Cycle Rides across Australia, showing the 63 relay stations

as well as the terrible stretch of sand along the Coorong coast of South Australia. Colonel Bridges endorsed the relay, which was subsequently known as the Dunlop Military Dispatch Cycle Ride. The military requested a report and a copy of the company's Coorong map, prepared especially for the ride. It was by far the most detailed yet drawn up of the isolated area, and provided the basis for all military and civilian road maps for many years. For Dunlop, the endorsement imparted an aura of national importance to the venture, a fact they highlighted in advertising.[5]

Sixty four relay teams covered the 1,143 miles in 69 hours and 35 minutes, an average of 16.4 miles per hour. It was 10 hours faster than originally conceived. Remarkably, in the Coorong, two riders pedaled, dragged and carried their machines over 56 miles of sand at an average of 12.2 miles per hour. In London, the relay was reported in *The Times*' Military Intelligence section. In Australia, Colonel Bridges referred to an important service having been done, "without any expense to the Commonwealth." The Minister for Defense noted that the matter had been discussed by the Military Board, but gave no further details. The comments committed the military to nothing and led to no large orders for cycle tires. The Australian military-industrial complex had not yet begun to roll.[6]

By mid-1911, the increasing speed and reliability of motor vehicles led to suggestions that they could now negotiate the 1909 relay route far quicker than cyclists. In response, Dunlop organized a second relay race for April 1912. The trusty bicycle was pitted against the far more fascinating, less reliable, but faster and more spectacular motor cars and motorcycles. The three groups of vehicles were handicapped, the bicyclists starting first.[7]

The race was held in atrocious conditions, as a storm front followed the riders eastwards. Nonetheless, the 64 teams of bicyclists bettered their 1909 good-weather ride by three minutes. The route was broken into four sections for the motor cars, with a pair of vehicles negotiating each leg together. They suffered no significant mechanical troubles, and in the early stages through the sandy Coorong astonished everyone by averaging 35miles per hour over the rain-compacted sand. They completed the journey

Two Dunlop Relay teams

in 46 hours, 44 minutes, for an average of 24.4 miles per hour. The motor cyclists rode 26 sections, in pairs, and fared the worst. On the terrible road surfaces, three riders suffered serious injuries from falls, and several machines had major mechanical problems. In the Coorong, the motorcycles managed one 36-mile stretch only one minute faster than the cyclists. Overall, they averaged 22.2 miles per hour.

A report on the relay race was filed by two Australian military officers. Since the relay ride required several weeks of organization, involved some 1,500 postal and telegraph communications, and would only be needed if those very facilities were inoperative, they rightfully concluded that "it would appear difficult, if not impossible, to organize such a ride in War." They also judged the bicycles the most reliable, and least likely to be put out of action by road, weather, mechanical or tire troubles.

The 25th City of London Cyclist Battalion, The London Regiment, Rye camp, 1911. Riders are linked together to pull the machine gun (left) and field cannon (right)

Amazingly, the officers deemed the problem-plagued motorcycles less likely to be "rendered useless" than motor cars, and adjudged them superior for messenger service, despite the fact that the cars were much faster and had no significant mechanical problems. It was a strange and unwarranted conclusion.[8]

After the turn of the century, many countries kept a close eye on one another's military activities. In the most extreme case, from half way around the world, the Australians published Notes on the Belgian Army. Interestingly, neither the cavalry nor the artillery received anything like the attention paid to the formation and training of the Belgian cyclists.[9]

When it came to the production of manuals, however, no one was as remotely prolific as the British. As early as 1889, English military cyclist drill manuals had been written. One was

Cycle Manuals

accorded a formal review in a leading magazine. The author, an Oxford Professor, probably wished it had not been. In the eyes of *The Saturday Review*, his fatal fault was sloppiness. "We cannot resist expressing great pity for Second Lieutenant (otherwise Professor) J. Cook Wilson, for his efforts in writing this *Manual of Cycling Drill*. But if our pity for him is great, it is far greater for those unfortunate cyclists who drill under his direction." Professor Wilson confused the meaning of "section;" called a "distance" an "interval;" classified a man as a "single file;" substituted left for right (and vice versa); halted men who were already halted; and—ultimate sin—confused artillery and infantry to the extent of calling a "rifle" a "gun."[10]

While the more obvious errors were gone from most manuals by the early 1900s, they continued to proliferate. The British published and re-published their *Cyclist Training* manuals and *Instructions for the Care and Preservation of Military Bicycles*. They prepared handbooks on many foreign armies, including details about their cyclist complements. They went so far as to note that Switzerland's 14 cyclist units, with a combined 2,590 riders, had a minimum height requirement of only five feet and one fifth of an inch. There seemed no end in sight, the War Office encouraged, perhaps, by the fact that *The Saturday Review* by then had more interesting things to review.[11]

In America, the first two military cyclist drill manuals, privately published in 1892 by Albert Pope, were of little practical utility. One, by Lieutenant William T. May, devoted seven of its forty eight pages to bugle calls. The other, by General Albert Ordway, was good only for "street and other performances," in the words of a military reviewer. In 1898, however, Captain Howard A. Giddings, of the Connecticut National Guard, executed a tour de force when he produced his 15-page *Manual for Cyclists for the Use of the Regular Army, Organized Militia, and Volunteer Troops of the United States*. It was the most thorough of its time. He detailed standard military procedures for mounting, dismounting, grounding, stacking and supporting cycles. Captain Giddings also discussed cycle maintenance; how to conceal them, ford streams, and scale walls with them; how to train cyclists; how to use the pistol when pedaling; and provided valuable information for cyclists engaged in military photography, raids, convoy escort, demolition work, map reading, signaling, reconnaissance, hospital service and on patrol.[12]

In the end, the numerous military cycling manuals and reports on foreign armies demonstrated one point very clearly: Tens upon tens of thousands of soldiers in England, Germany, France, Sweden, Norway, Switzerland, Austria, Italy, and Estonia were mounted on bicycles, ready to pedal hither and yon.[13]

But What Role?

The critical aspect of all the official manuals was what was not said. Among the innumerable pages could be found no thoughtful analysis of the possible role of the bicycle in future warfare; what was fundamentally different about it; or how it might alter combat, tactics, strategy or intelligence gathering They seemed to treat the cyclists, for the most part, as wheeled pedestrians or cavalry substitutes. Assessments of the Boer War experience, the only effective military use of the cycle to date, were almost totally lacking. The official Boer War histories did not even refer to cycle troop activities, and only minor testimony on bicycle use was elicited during the Boer War inquiry.

Among the few to attempt to draw military cycling lessons from the war was Major-General Frederick Maurice. At a speech at the United Service Institution in May 1901, in London, he cited "the importance of mobility" which the Boer War had demonstrated. While he did not go so far as Arthur Conan Doyle, a wartime observer who argued that the cavalry should be abolished altogether, Maurice argued that "in certain circumstances there was no means of transferring forces that was superior to the bicycle."[14]

Like others in that era, however, Maurice was never able to come to grips with a fundamental problem—how to communicate with fast moving bodies of cyclists. That issue became painfully clear at the Sussex Maneuvers of August 4-6, 1900. In that exercise, military cyclists were to resist an invading force on the Sussex Downs, by riding rapidly south from metropolitan London. As one observer judiciously noted later, Major-General Sir Frederick Maurice, who was in charge, "should be congratulated" for trying to coordinate numerous dispersed cyclists strung out along countless parallel roads. In fact, it was a shambles.[15]

In 1906, further maneuvers confirmed that if bicyclists were to be used effectively in large numbers, they would have to be formed and drilled in integrated units. In the pre-walkie-talkie age, command over isolated, rapidly moving, individual groups of riders was effectively impossible. Consequently, when the Territorial Force was formed in 1908, 10 separate cyclist battalions were created. By the time war broke out in August 1914, there were over a dozen.[16]

Between 1901-1914, there were only scattered reports of military cycle usage in wartime conditions. The Japanese apparently employed cycle messengers in the Russo-Japanese War, but it is not clear how extensive that use was. Some Serbian, Bulgarian and Greek soldiers apparently used bicycles in the Balkan Wars against Turkey in 1912 and 1913, according to fragmented accounts. As well, cycles were "in evidence" during the Italian campaign against the Turks in the Tripoli hinterlands. Aside from these minor instances, the period between the Boer and First World Wars was essentially an era of speculation as to the most effective use of cyclist soldiers.[17]

The Ironclad and the Iron Steeds

The tank made its first appearance in 1916, on the western front, in an effort to break the stalemate of trench warfare. Its antecedents, however, went back many decades. The incentives to create an effective tank-like device increased as the American Civil War, the Franco-Prussian War, and the Boer War showed the difficulty of trying to advance unprotected soldiers against the sophisticated firepower becoming available. The advent of petrol engines enhanced locomotive possibilities, and a critical development, the caterpillar track, was finally perfected and in use on agricultural machinery prior to World War I.

While the tank was clearly the product of many minds and, it could be argued, would have eventuated anyway, H. G. Wells' article, "The Land Ironclads," is considered germinal. The thirty-three page story, published in *The Strand Magazine* in December 1903, is a mandatory reference for writers of tank histories. It was yet another of his outstanding expositions of new technological ideas in fictional form. Wells was truly astonishing in his foresight. He developed the theme of aerial warfare in a series of writings from 1898 through 1907. At a time before most military men even thought of the plane as a potential weapon, Wells realized its implications in converting warfare from fronts to much larger areas, in which "all directions lead everywhere." Finally, in 1913, he combined the airplane with the infant scientific work on radioactivity to conceive of future airborne atomic warfare.[18]

One thing rarely remembered is that Wells foresaw cyclists as the principal supporting forces for his ironclads. In his vision of events, after the tanks rolled past the enemy trenches, the defeated soldiers lined up alongside the machines and waited for cyclists to take them prisoner. In one scenario, Wells portrayed the cavalry as unsuccessfully trying to ride down the cyclist troops. However, the cyclists retreated before the horsemen "with a sufficient margin of speed to allow of frequent dismounts and much terribly effective sharpshooting." It was a lesson straight out of the Boer War. In defeat, the cavalry suffered the ignominy of handing over their horses to Wells' wheelmen.

Military cycling trials at the West Australian Cricket Ground

H. G. Wells's War of the Words

Within Wells' fertile mind, the coupling of cyclists with the land ironclads was not momentary serendipity, but was based upon much thought devoted to the role of the bicycle over many years. In 1898, for example, he wrote a perceptive story about the effect of the machine in breaking down social class barriers in England, appropriately titled *The Wheels of Chance*. It described men and women of various classes cycling far beyond their normal social haunts, unconstrained by chaperones and usual patterns of behavior. Few others assessed the impact of the bicycle so cogently, creatively and with such a penetrating mind.

In late 1900, Eustace Balfour's *Cyclist Drill* manual came to Wells' attention. It had first been produced in 1889, and republished twice since. Wells was appalled by the "little pink book, rather badly printed," which was "the most remarkable piece of reading I have happened upon this year." He penned his reactions in *The Fortnightly Review*, in December 1900. The drill manual opened with the incomprehensible and unexplained quote "(Wt. 13,016. 10,000. 9/00-H & S. 4794)." Wells henceforth referred to the anonymous author as "Wt. 13,016," and warned that his comments would be, on the whole, "unflattering."[19]

It was British understatement, par excellence. The 15-page critique was devastating. Wells characterized the unacknowledged compilers of this little work as having "neither the intelligence nor the imagination necessary for their task." He focused upon the crucial issue: the authors simply did not appreciate the unique nature of the bicycle. They had merely adapted quite obsolete infantry drill, developed originally out of the needs of pikemen, to cyclists. As Wells concluded, "that a cyclist is after all a cyclist, has clearly never entered their heads." He was absolutely merciless in attacking what he deemed to be trivial, misguided, and inappropriate concerns of the manual writers: among them, how to handle officers' swords; where to tie the great coat; how to use cape-straps ("whatever cape-straps may be"); and the incomprehensible fact that a fighting cyclist's valise, containing all his field items, would be carried in a separate cart that could not remotely keep up with him. He poured scorn upon the incessant concern with drills.

But Wells went beyond literary destruction. He spent many pages discussing his views on what was needed for future cycle warfare, ranging from the mechanical characteristics of bicycles, to tactical maneuvers. He also proposed a theoretical battle between cyclist soldiers using tactics based upon recent Boer War experiences (which he criticized the manual writers for ignoring), and cyclist soldiers trained using the *Cyclist Drill* manual. As he concluded, there is "urgent need of a sane experimental handling of this problem of cycling tactics."

To Wells' utter astonishment ("it did not occur to me that any serious attempt to answer my criticisms would be made"), Lieutenant Colonel Eustace Balfour ("I have no wish to be prominent") came out of the War Office woodwork, identified himself as the author (with Colonel Lloyd), and wrote a 10-page reply to *The Fortnightly Review*. He criticized Wells' "foolish sarcasms," personally attacked him, and offered a review of military cycling history that was largely irrelevant to the issues at hand. Balfour's reply only reinforced Wells' opinion that the manual had been poorly thought through.[20]

Wells was justifiably taken to task on some minor points. As Balfour noted, if Wells could not see the necessity for drill at all, it "was proved by the attempt to work without it." Balfour also defended the valise carts on the basis that they were for carrying long-term campaign items, which the cyclist could do without for a short while in battle. However, with respect to the cape-strap matter, Balfour was unconvincing in his argument that "anyone who knew what a "rug-strap" was could deduce by analogy the inward meaning of the word "cape-strap"."

On the critical issue of cyclist tactics, the crux of Wells' review, Balfour was inept in response. As to Wells' suggestions about the tactical use of bicycles, Balfour hardly said anything. His illogical reasoning was that while he had intended to deal with Mr. Wells' "constructive method of tactics at some length," it was "not worth while to do so, the reason being that it is not a method at all."

Balfour abruptly finished his response to Wells' criticisms because "it is mere waste of time to show up any further a man who approaches Military Cycling in the spirit which he exhibits, and who has clearly never read one word of the literature on the subject, English or foreign, nor ever personally taken the trouble to watch any actual cycling manoeuvres." Wrong on all accounts,

Balfour went on to add: "Poor Mr. Wells. He must now sit revealed as the armchair ignorant person, posing as one who has but imaginative knowledge."

Wells' brief, tight three page reply followed in the next issue of *The Fortnightly Review*. "In an issue of such grave public importance," he wrote, he was forced, "quite against my inclination, to deal in a hostile and destructive spirit—no longer with the Unknown Entity responsible for that Drill-Book, but with that Entity revealed—as an interesting (and quite naturally extremely irritated) individual." Wells ignored trivia and immediately attacked the fact, disclosed by Balfour, that there had been no substantive modifications to the original 1889 drill book, despite two reprintings and many intervening years of cycle maneuvers. Wells believed that that fact alone clearly demonstrated the need for drastic War Office reform. It had "serenely printed and put forth the twelve-year-old projects, practically unmodified, of a man who boasts of forming conceptions of cycling on an old-fashioned ordinary." Wells was particularly angered by the fact that Balfour had acknowledged that there had been some worthwhile lessons, but felt that he could hardly ask the Adjutant-General to re-write the drill book to incorporate them. This apparent unwillingness to make germane changes, coupled with a lack of creative thinking, was among those things Wells had long criticized about the British military. Balfour was bearing the brunt of far deeper concerns when Wells concluded that "by virtue of this harmless-looking little pink pamphlet, unless some grievous outrage to Lt.-Col. Balfour's pride of authorship is speedily done, I am convinced that men of my blood and class will be brought to intolerable hardships, to shame and surrender, to useless struggles, and wounds and death."[21]

In a few years, in the fields of Flanders, Wells' tanks, bicycles, and forebodings would come together.

4. World War I: The Allied View

"There was never a greater tragedy than World War I. It engulfed an age, and conditioned the times that followed. It contaminated every ideal for which it was waged, threw up waste and horror worse than all the evils it sought to avert, and it left legacies of staunchness and savagery equal to any which have bewildered men about their purpose on earth."
BILL GAMMAGE, THE BROKEN YEARS, 1975

"Our fate was to be used as reinforcements to infantry battalions, and the original unit evaporated."
E. C. BOND, PERSONAL LETTER RE: CITY OF LONDON CYCLIST COMPANY, 1963

The Great War started in August 1914. Military plans envisioned the massive use of cavalry, with millions of animals and men being moved around on railway networks to face one another. However, the internal combustion engine, machine gun and artillery quickly altered the dimensions of the war. The British Expeditionary Force took only 842 motorized vehicles to France in August 1914, 90 per cent of those requisitioned. Four years later they alone were using some 113,000 cars, trucks, motorcycles and motorized bicycles. Overhead, the airplane came of age, and assumed tactical and strategic importance. By the end of the war more than 200,000 had been built, and the Germans were bombing London from the Continent. Ultimately, a new technology—the tank—helped to break the stalemate that the trench-based war became. Amidst it all, both sides also used hundreds of thousands of bicycles. No one knows exactly how many.[1]

Initially, it was believed by some that the war would be over by Christmas, with soldiers pedaling speedily to Berlin—or Paris, depending upon the perspective. Unfortunately, bicyclists, like the cavalry and infantry, encountered the machine gun and fearsome artillery. The machine gun was no longer the heavy, cumbersome device drawn behind a horse carriage that it was in the Boer War, and proved to be far more than just a rifle that could fire faster.

A British "Tommy" examines a captured German bicycle with spiral spring tyres. A thin strip of hardened steel rested on springs fastened into the wheel rim. An article in The New York Times *in July 1918 said that the tires were invented in Sweden and used in Germany in response to a wartime shortage of rubber. A Norwegian newspaper said that "the invention proved to be so simple and adaptable that the wonder is that it was not thought of long ago." It had been. In 1897 Lt. James Moss's troops tried a variation during their American overland ride, for example. The tyres proved impractical and were no match for pneumatic tyres.*

The relatively light weapon, with a well-trained crew, could be readily moved and set up, and was capable of virtually annihilating a field of advancing troops. As well, the artillery of that era was of a magnitude never before seen on the battlefield. Firing huge shells from immense distances, with remarkable accuracy, it too could substantially eliminate entire masses of advancing troops. Railway networks and communications systems facilitated the coordinated transport of large numbers of soldiers from one point to another, and enabled them to be kept supplied with food and weapons. The result was that millions of men and materiel from each side converged to do battle on a scale never before seen. In the ensuing contests of flesh against artillery and machine gun bullets, both armies sought protection.

They found it by digging. Each side built thousands of miles of interconnecting zigzag trenches, and strung barbed wire between themselves, sometimes only a few yards apart. The troglodytic soldiers sat in water and mud and cold and heat in their trenches. When ordered, they arose from the earth and jogged, walked, ran, or crawled across no man's land, over ground churned up by artillery, and sometimes mined. They moved toward the enemy and their machine guns, beneath the falling artillery shells, and were killed and wounded in staggering numbers, up to 80,000 men in a single day. The machine gunners, who dealt so much death so fast, were primary targets for the charging soldiers and their supporting artillery. Consequently, the average life of a machine gun crew in battle was less than 30 minutes. Ultimately, no one was safe, and there were no winners. When the charges were over, the soldiers retreated to their trenches, recovered what dead and wounded they could, and sat "eye-deep in hell," listening to those slowly and agonizingly dying in no man's land. It went on for four years.[2]

At the outbreak of war, cyclist soldiers demonstrated their great mobility and value. On the fluid battlefields during the first weeks of fighting, cyclists were everywhere, often ahead of all other forces, taking important positions, bridges, and roads. Once trench warfare set in, however, bicycle troops were relegated to use behind the lines for such things as delivering messages, moving troops, directing traffic, and generating electrical power. The device was a

Dutch soldiers during mobilisation for World War I

truly valuable tool. But no man's land was no place for a cyclist, and trench warfare was not conducive to cycle charges.

Early Optimism

When the war started, there were numerous organized cycle units on both sides, and great anticipation of using them. Many officers, Allied and German, had studied the terrain from bicycles. The Commander of Britain's Staff College, Brigadier General Henry Wilson, for example, pedaled along the French, Belgian and German frontiers in 1909. He returned in subsequent years for yet more cycling surveys of the area, becoming intimately acquainted with the distances involved, and rates of travel possible on the two-wheeled machines.[3]

In 1914, the British reportedly had some 14,000 men in formal cyclist regiments and battalions. By war's end, there were over 20,000. That did not include an unknown, but very large, number of cyclists attached to virtually every British military unit (including cavalry) as messengers, orderlies and couriers. As well, from early 1916 there was a reorganization that led to mounted troops being transferred from the control of the divisions to the corps; each corps henceforth included some three cyclist companies, totaling 500 men. One writer calculated that at least 100,000 British soldiers were mounted on bicycles in some capacity, and

British cyclists riding through a destroyed village towards the trenches

another source suggested that at least 150,000 French and Belgians rode cycles.[4]

The Manual of Military Law formally recognized the machine in warfare. According to opinions issued at the 1907 Hague Conference, bicycles could be impressed for military use, but were classed as "personal belongings" that remained the property of prisoners of war. The importance attached to the machines can be measured by the fact that when an English volunteer had his bicycle stolen while attending his physical, it was the subject of an editorial in *The Times* of London.[5]

In the first few weeks, before the war became trench-bound, there were numerous articles and personal accounts relating the exploits of cycling soldiers. They variously rode down and captured the enemy, saved their comrades's lives, and pedaled through hostile fire to deliver messages and warnings. *The Bicycling World and Motorcycle Review*'s titles for its articles --"English Cycle Rider Saves French Column;" "With the Cyclists in the Fields of Battle: An English Correspondent Narrates some Exciting Experiences with and Between the Opposing Armies;" and "Despatch-Riders Give Thrilling Accounts of Their Adventures "Mid Shot and Shell""—typify the tone and content of many such writings.

The types of bicycles in use varied widely. The British were generally equipped with official military bicycles, which were solid and reliable, but described by one rider as "anything but a speed iron." Most were reportedly fitted out with Sturmey Archer three speed hubs. As to the French Army, some writers suggested that Captain Gerrard's folding model was universal. However, photographs suggest that that was not the case. Indeed, pictures of General Joffre reviewing a group of alpine soldier-cyclists reveal a variety of bicycles, including standard touring models.[6]

The bicycle's major advantages were highlighted in the early reports. The machines proved to be quiet, and could move with facility along roads crowded with wagons, cars, large numbers of marching troops, and fleeing civilians. Soldiers readily pedaled 50 to 100 miles per day, even in very hilly country, and night rides of 50 miles were common. Laden German ammunition wagons "had no chance of escape from cyclists" and were captured and destroyed by them, a fact reported in the allied press

A cycle battalion, including Australians and Canadians, parading before Lord French

and later lamented in a German analysis of wartime cycle use. A British soldier, G. W. Holderness, found that "the push-bike could be made to go almost anywhere and at a speed which left infantryman far behind." A Belgian cavalry officer, Lt. Raoul Daufresne, admitted that the bicycle was much quicker, no small concession from a man who had been complimented by the King of England for his riding at London horse shows. The advantage of silence was well demonstrated by a group of allied cyclists who heard approaching motorcycles from afar, dismounted, hid alongside the rode, and killed the two German messengers as they passed.[7]

During the retreat of the British Army from Mons to the Marne, in mid-August, French and British cyclists played major rearguard roles. Prone soldiers, with their bicycles lying alongside, were relatively inconspicuous in the trees, foliage and grass. Later, at the battle of La Cateau, southwest of Mons, one group of French cyclists folded their machines, strapped them on their backs, and charged into the face of the enemy's rifle and machine

gun fire. "At every rush cyclists fell moaning or motionless" among the beet roots and pools of mud. It was followed by the field guns "bespattering the prone figures." Eventually the survivors unfolded their bicycles and retreated on them. One soldier found that his hinge would not unflex, and gave himself up for lost. Another soldier quickly helped him get the machine unfolded, however, and he managed to pedal away to safety.[8]

Bicycles were everywhere during the war. At the invasion of Gallipoli, that rugged peninsula where British Empire troops faced off against the Turks, the machines were optimistically landed on the beaches of Cape Helles, only to be abandoned, unused, when the British evacuated. Near the end of the war, in Mesopotamia, a soldier who was an English Cycle Touring Club member photographed bicycles upended under cavalrymen's tents; for many tasks, they were quicker and easier to use than saddling a horse. Even the American Expeditionary Forces acknowledged the machine's value. When they sailed to Europe,

Indian cyclists at a crossroads on the Fricourt-Mametz road at the battle of the Somme, July 1916. They were part of that government's contribution of $500 million and many men to the allied cause, having been promised increased post-war autonomy. After losing the cash and many lives, India received tightened British rule.

in 1917, the Yanks took 29,000 bicycles with them.[9]

Trench warfare, however, proved inhospitable to specialized bicycle fighting units, and bicycles were relegated to messenger work and behind-the-lines activities. Soldiers from cycle units were routinely reassigned to infantry units to serve in the firing line. The result was that many cycle units were effectively disbanded, or operated only at very limited levels. That fact was driven home pointedly in the case of the London Cyclist Battalion, the most famous in the world. Through late 1915 they rode coastal patrols in the south of England, watching out for a possible German invasion and fighting fires set by raiding German aircraft. When they finally shipped abroad, to India, in December 1915, they went as infantrymen, without bicycles. "It was a bitter blow," wrote one battalion member. It was all the more so when, in mid-1916, Indian troops who were brought to France used bicycles.[10]

The Anzac Cyclist Battalions[11]

The life of cyclist soldiers on both sides during the Great War is well illustrated by the Anzac (Australian and New Zealand Army Corps) Cyclist Battalions, which served on the Western Front. The 1st Anzac Cyclist Battalion was comprised solely of Australians. The 2nd Anzac Cyclist Battalion (widely known as the N. Z. Cyclist Battalion) was a Kiwi unit. The two battalions spent the war moving frequently from one billet to another along the Western Front. The New Zealanders, for example, lived in 82 different locations in their 32 months of action. Many soldiers were temporarily reassigned to other units, under the command of other officers, and at any given time either battalion might be half-depleted. Those remaining were engaged in such tasks as directing traffic, unloading railway goods wagons, felling trees, burying cables in no man's land (a New Zealand specialty), harvesting hops for local families, repairing trenches, and burying the dead.

There would have been few units better guaranteeing military anonymity than the cyclist battalions. The routinely-understrength Australian battalion got scant reference in that country's *Official History* of the war, and virtually no mention in any other book. The Kiwis were slightly better known, principally as a result of a *Regimental History* the officers produced in 1922. In essence, the Anzac Cyclist Battalions were the wrong thing for the wrong place at the wrong time.

However unheralded, the Anzac cyclists carried their share of the war burden. Of the 708 men who served in the Kiwi battalion at one time or another, 59 were killed and 259 wounded (51 more than once). They took home—healthy, crippled or posthumously—72 British, French and Belgian medals. The unit also received a flag from the town of Epernay for its role in stopping the Germans just short of that community during the Second Battle of the Marne. The Australian battalion was luckier. Working a much greater portion of its time behind the lines in traffic control and other functions, only 13 members of the battalion were killed while serving with it. But the number of 1st Anzac Cyclist Battalion members who were killed or wounded after being transferred to other units was very high. Of the 295,000 Australians who served in France, one fifth died in service, and another 52 per cent were casualties.

After joining the Australian Imperial Force, Jack Hindhaugh was evacuated from Gallipoli to Egypt in 1915, with wounds. There, the Anzac forces were doubled during January and February 1916. It meant much shuffling of troops between

Captain Jack Hindhaugh (second from right) with fellow officers in Egypt

"Gentleman Jockey Jack" Hindhaugh at the Cairo Races

units, and decisions as to who would stay in Palestine and who would go to the Western Front. During that period, Jack Hindhaugh, then a Captain, was an aide-de-camp to Colonel (later Lt.-General Sir Harry) Chauvel, Commander of the Desert Mounted Corps. Hindhaugh was an excellent speaker, socially confident, ruggedly handsome, physically tough, and a noted horseman. He broke in mounts for other officers, conducted riding tests, played polo, and rode in races in Cairo. He was known as "Gentleman Jockey Jack."

When the Anzac forces were restructured, Captain Hindhaugh sought command of a unit. On March 6 he was offered a camel squadron, but turned it down: "camels stank." Hindhaugh was subsequently given the command of a Cyclist Corps. He made no comment at the time in his personal diary as to his reaction. However, two months later, after General Birdwood had inspected his troops, Hindhaugh wrote that "Birdie asked me if I had come down to riding cycles and could I ride them as well as I can a horse."

Why the Anzac Cyclist Corps were formed is not entirely clear, and it appears to have been a strange commitment of men. During the previous two decades Australia's military authorities had refused to sanction independent cyclist units. The Chief of the General Staff, Colonel (later Maj.-General Sir William) Bridges had long argued that cyclists were best employed individually or in small groups as messengers for larger units. As well, the fighting conditions on the Western Front were quite clearly defined by early 1916. It was not a war in which cycle units, per se, could play a significant role. Most likely the Cyclist Corps came about because the Australians based their new divisional structure upon the British pattern, which included cyclist units. Generals White and Birdwood, who expanded the Anzac divisions from three to six in only a few weeks, had little time for reflection—if the British were doing it, it was good enough for the moment.

Captain Hindhaugh, in turn, had even less time to prepare his men. He was advised of his command on March 14th; on the 15th he got a cyclist's drill manual from the British; on the 17th he received orders to move his group to Alexandria; and 10 days later they shipped out to Europe. Understandably, there were problems in fitting out and training the unit in such a short time. Non-military bicycles were pressed into service, and puncture kits, accessories and spare parts were in short order. Some men still did not have machines by embarkation. Amazingly, "a number of the men had never ridden a bicycle previously, and a good deal of time was wasted in teaching them." In the brief period available, they did not become proficient riders—they could not even win the bicycle races at a divisional sports day. Far worse, however, was the fact that some of them were firing rifles for the first time. Such was the preparation of Australia's 1st Anzac Cyclist Battalion when it shipped out to Europe on March 23rd.

New Zealand officers of the 2nd Anzac Cyclist Battalion

The New Zealand cyclist company, in contrast, was formed in New Zealand, had two months to recruit men (principally from the Mounted Rifles), and made sure they could pedal and shoot. In the southern hemisphere autumn of May 1916, the unit sailed for Europe with 204 men and 206 bicycles. After 46 days of travel, they arrived in Egypt, suffering from seasickness and the terrific heat of the Gulf of Suez. Soon afterward they left for France.

By early May the various Australian cycle units had been reorganized into the 1st Anzac Cyclist Battalion, with over 300 men. Hindhaugh wanted command, but did not get it. A rapid series of command changes left discipline in a poor state. Hindhaugh cited one company as having "as usual one officer on parade—they take it in turns to get full every night. Damned shame." Disgusted, he asked to be relieved of second in command. To his surprise, Hindhaugh was placed in charge of the battalion. To celebrate his appointment and new rank that night, he went to the Officer's Club in the nearby town of Poperinghe with his mates. They had "a good time." So much so that one of his fellow officers "had a few falls off his bike on consequence on the way home."

Cycle ordnance workers under shellfire, Etrelliers, April 25, 1917

Coming from relatively sunny climes, the Aussies found the cold winters hard to take. They discovered that riding bicycles through deep snow—a new experience—was impossible, and pushing the heavily laden machines up frozen, snow-covered slopes was taxing. On one especially slippery uphill climb, "spills were frequent and our tempers sullen." The colonials were not alone in their despair. An Englishman in Flanders found that "a whole winter in this war-stricken country, with its atrocious roads and impossible by-ways is enough to dispel any ideas [about the joy] of riding." They also suffered the occasional problem of having to pace themselves with accompanying slow horse transport. During one night's ride between bivouacs the troop and transport traffic was so heavy that not only was progress painfully slow for the cyclists, but one New Zealander had his leg broken when run over in the dark by a motor lorry.[12]

The New Zealand cyclist troops were eventually used to lay cables between the front line trenches, the primary support trenches (about 50 to 300 yards back), and the subsidiary support trenches (another 700 yards or so back). It involved digging ditches seven feet deep, to protect the cables from shelling, laying the lines, and re-covering them. Standing sometimes waist-deep in water in shell-pocked terrain, it was common for partially completed ditches to collapse, and have to be re-dug. It was normally done at night, without lights, to avoid the enemy pinpointing the workers and calling artillery down upon them. The New Zealanders proved so adept that most were eventually assigned to supervisory roles.

On several occasions Major Hindhaugh was excited by the prospect of his battalion becoming directly involved in the fighting as a unit, but in each case he suffered frustration and

disappointment. A combination of monotony, dashed hopes for combat, and the dragging on of the war took its toll on morale. As well, diphtheria and meningitis epidemics forced an isolation of the battalion at one stage, influenza swept the ranks, and foot problems were endemic. Bill Gammage, in his superb study of Australian soldiers in the war, noted that by early 1917 "the buoyancy of even a year earlier was now foreign to them." Major Hindhaugh was no exception. In April 1917, when all troops in the area but his had moved out, he commented that "the village is very quiet. We are left in a backwash—hope they forget us altogether." In essence, the battalion suffered the stresses of war, but without the cathartic sense that it was contributing directly to its resolution. In October 1918, Major Hindhaugh left for a furlough in Australia. It is not clear whether he was to return to Europe if the war continued. In the event, it did not, and he did not. At his farewell, on a cold day at La Mesnil, the greatcoated commander of the 1st Anzac Cyclist Battalion wore spurs. His diary suggests no remorse or regret at leaving. Like most by then, he was "fed up", "sick up", and "tired."

Early the following year the 1st Anzac Cyclist Battalion was disbanded and consigned to history. Among its last official acts, some of the men paraded before the King, on Salisbury Plains, in England. As the Australians pushed their machines past him, a few were caught on motion picture film for a few seconds.

Military bicycle units were not particularly photogenic, with one exception—the Italian Bersaglieri. They were distinguished by their broad-brimmed hats, reminiscent of those worn by British colonial troops, but with large flowing plumes attached. *The Illustrated War News* of England described them as "splendid fighting men. On the march they swing along with a *verve* and impetuosity that is only equalled by the best French infantry." The magazine appeared entranced by them, and featured them several times during the war.[13]

The Bersaglieri

In the field the Bersaglieri in fact wore either steel shrapnel helmets or soft cloth peaked caps. They were specially chosen, solidly built, and strong, since much of their work was carried out in the Alps, and they had to be able to carry both their field pack

Italian Bersaglieri on the march

and 35 pound folding Bianchi machines over rough ground and up steep slopes. An English observer categorized them as rugged, heavy mountaineers. Their dull gray-green bicycles had spring frames, fixed wheels, rat trap pedals, and a rim brake on the front wheel. Astonishingly, according to one account, the soldiers used

Italian Bersaglieri with folding Bianchi bicycles

cushion rubber tires, with only the officers having pneumatics and freewheels.[14]

In World War I the Bersaglieri served as special light infantry troops attached to each Italian Army Corps. In 1918, fast riding Bersaglieri units were among the first to re-occupy Venetia as the Austro-Hungarians evacuated it. But their success was mixed. They suffered a particularly bad setback in 1916, at Casara Meata, when they were among the last reserves to be thrown against the advancing enemy, and were all captured, along with their cycles.

Bicycles and Bombers

During the First World War London was subjected to numerous German aerial bombardments. They began in May 1915, with Zeppelin airships dropping over 100 bombs on Southend. By 1917 the Germans were using airplanes. On June 13th that year, 14 Gothas attacked in broad daylight, killing 162 Londoners, and injuring 432 others. Among them were 46 children in one school. By the war's end, 1,413 people had died and 3,407 were injured by the attacks. Authorities quickly developed plans to warn of the impending air raids. As soon as the enemy craft were spotted along the coast, the information was telephoned to London. There, motor vehicles and bicyclists traveled about neighborhoods, alerting residents to the danger. Cars had signs on them with such phrases as "Take Cover", and cycling constabulary pedaled around with messages hanging from their necks, blowing whistles and ringing bells.[15]

A British South African Police officer with General Northey's force, on patrol in Northern Rhodesia during World War I, crossing a crude bridge

In The End ...

At the conclusion of the war, the mundane bicycle could be said to have acquitted itself well. It proved to be a valuable tool that supplemented and facilitated the efforts of virtually every unit. Among its uses, however, there was little that was surprising, and nothing that was spectacular, and in the final analysis it was totally overshadowed by the astonishing rate of adoption of the wide variety of mechanized transport. As a result, one observer noted, "the general newspaper correspondent does not give the bicycle the amount of credit it deserves."[16]

Nor did many in the military. Over the next few years, in many countries, cycle units would disappear, one after another. Such was the fate of the New Zealand 2nd Anzac Cyclist Battalion, which was disbanded in 1919 and never resurrected. In the *Regimental History*, written in 1922, in the last sentence, the officers took consolation from the fact that everyone was "proud to have formed a part in the Great Army which rid the world of a Prussian menace …"

5. Die Radfahrtruppe: 1914-1945

"... persuade Monty [General Montgomery] to get on his bicycle and start moving."
                GENERAL DWIGHT EISENHOWER TO WINSTON CHURCHILL, 1944.

In 1945 the U.S. War Department noted that the German Army of 1939 was "a model of efficiency, the best product of the concentrated military genius of the most scientifically military of nations." Whether it was an accurate assessment or not (there are many who would dispute it), the foundations for that model were derived partly from careful analysis of First World War experiences. Among the analyses was *Die Radfahrtruppe*, a systematic assessment of the bicycle completed in the 1920s. The Teutonically thorough study of the machine's wartime role highlighted the strengths and weaknesses of cyclists and cycle units in various circumstances. The Germans deemed many of those circumstances still valid when preparing for World War II. The blitzkrieg onslaught that they unleashed against their neighbors, beginning in 1939, saw panzer tanks and stuka dive bombers work together to demoralize and devastate opposing armies and conquer territory at a rate hitherto unknown in warfare. Soldiers on bicycles were not far behind.[1]

Post-World War I Perspectives

After World War I, views on the bicycle's military future varied widely, heavily influenced by the gasoline engine. In 1916 Henri Deterding, head of Royal Dutch Shell and the most powerful man in the international oil industry for 25 years, summarized it simply: This is a century of travel and it would be oil based. During World War I the British increased their motor vehicle usage 134 fold, for example, and in 1917 the Americans brought with them some 4,000 horsepower per division. The 1920s saw petroleum based transport increase at a remarkable rate. In that context, many saw little prospect for bicycle troops.

But the increasing reliance upon oil powered mobility was at a price—dependency upon oil. That lesson was driven home

Dutch troops in training, 1925

in England in 1916, which experienced petrol shortages as German submarine raids severely restricted oil shipments to the British Isles. In France, lack of fuel threatened that country's ability to carry on the war effort and Clemenceau appealed to Woodrow Wilson for help. "Gasolineless" Sundays became a fact of life for awhile in the United States, which supplied some 80 per cent of allied war needs. The allies also cut Germany off from all oil suppliers other than Rumania, which remained neutral until 1916. After Russia entered the war on the eastern front, Rumania declared war against Germany. General Erich Ludendorff appreciated the crucial nature of Rumania's oil supplies and sent German troops to capture the facilities. Simultaneously, under great pressure from the Allies, Rumania agreed to the voluntary destruction of its oil fields by a British team. By the time the Germans captured the

Dutch soldiers on field manervers, Winter of 1939-1940

Ploesti fields, the wells, tanks, and related facilities were ablaze or destroyed. It took the Germans five months to get the field back to one third of its prior output. Although they eventually achieved 80 per cent production by 1918, at one point they were nearly unable to continue the war. The Germans kept that lesson well in mind during the next two decades.[2]

In England, the bicycle appeared to have drawn its last significant military breath. The British formally disbanded the Army Cyclist Corps and all other cyclist units per se. By 1922 "cyclists as such had disappeared from the army", as they had three years before in Australia and New Zealand. Among the results were bargain sales of machines, including "1,000 pedal bicycles by the best makers" which were auctioned off with no reserve, in August 1919. The epitaph for cycle units in the British Army was essentially written by the members of the London Cyclist Battalion. Along with the Italian Bersaglieri, it was the most noted cycle unit in the world, "devoted entirely to the development of cyclist tactics." Originally formed in 1888 by the 26th Middlesex (Cyclist) Volunteer Corps, the battalion was demobilized in July 1919. In the unit's history its officers honestly and accurately concluded

Dutch cyclists crossing a river, Winter of 1939-1940

that the London Cyclists achieved no "outstanding military distinctions or inherited any great traditions." They went on to add: "Inspired by no special exploits in the Great War, it is, for a large part, the chronicle of what was, originally, an interesting military experiment, conceived, developed and abandoned well within the span of a normal adult lifetime."[3]

Despite Britain's quick abandonment of formal cycle units, the bicycle continued to be a valuable transport tool around military bases and on airfield tarmacs. War Office handbooks on cycles continued to be issued. One of them, after describing and defining the two basic models as Mk. IVB and Mk. IV*B, respectively, then inexplicably designated them as Mk. IVA and Mk. IV*A, once issued.[4]

In many other countries, in contrast, bicycles remained in widespread military use. In Italy, the War Ministry's 1927 cycle manual was 194 pages long, with 20 pages of illustrations of the various types and sizes in service, including folding models and some mounting machine-guns. A 1939 manual specifically dealt with the role of bicycles and small motor vehicles in moving troops and ammunition. Switzerland's post-war army increased the number of bicycles in use, and the Belgians, Dutch and French continued to employ large numbers of cyclists. Even in America, that bastion of motorized transport, the World War I experience led the army to equip infantry divisions and field artillery units with some bicycles. In Russia, in 1936, five officers of a frontier guard unit on the Ukrainian border completed a 269 day ride around the Soviet Union. They traveled from Kiev to the eastern Siberian provinces, and back to Moscow, via the Karakorum desert in central Asia and the Chinese frontier. It was an amazing journey through widely varying terrain and climatic conditions.[5]

By 1935 a number of British officers had second thoughts about the abandonment of cycle units. In the *Journal of the Royal United Service Institution*, Captain R. L. K. Allen, of the Royal Welsh Fusiliers, reiterated many of the standard arguments for using bicycles, reviewed their use in World War I, and concluded that they would greatly enhance the mobility and flexibility of infantry. He noted that European nations still "appreciated their utility", and suggested that the English should consider re-introducing them. Allen acknowledged that officers commanding cycle units would be a problem, however, as they preferred to use motor cars. He suggested that they be issued horses. In a subsequent letter to the editor, Brevet Major Bruce, based in the Sudan, argued against inclusion of bicycles in infantry units. In closing, he snidely suggested that future cycle unit commanders should ride in sidecars, with their "bat-men doing the footwork." It was the 1890s all over again.[6]

Die Radfahrtruppe

At the outbreak of World War I, the Germans learned some quick lessons about the machine's value. In a two week period from late September 1914, five major raids, including 16 demolitions, were carried out behind German lines by Belgian cycle detachments, each with about a 100 soldiers and a small demolitions party. The Germans rapidly expanded their small cyclist units to include entire battalions of pedalers. Like the allies, they pressed a hodgepodge of models into service and took advantage of the cyclist's major asset: speed. As Barbara Tuchman noted, the waves of German infantry, artillery and staff officers in automobiles that swept into Liege and Ghent were preceded by "companies of cyclists who sped ahead to seize road crossings and farmhouses and lay telephone wires."[7]

In 1925, Major Rudolf Thiess produced *Die Radfahrtruppe*, in which he surveyed the bicycling experiences of both sides in World War I. It "shed much light on the versatile and successful use of the cyclist units, especially in mobile warfare." The Major offered suggestions and recommendations as to their future employment.[8]

As messengers, cyclists played a major communication role, and were invaluable where telephone connections did not exist. They combined effectively with cavalry to operate out of central radio locations, pedaling about surrounding areas to deliver information and orders. Thiess suggested that in future battles two cyclists should be sent via different routes to try and assure delivery of a message; in non-combat zones, one was sufficient. In particular, marching columns should clear the best side of a high-

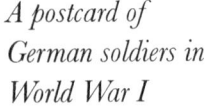
A postcard of German soldiers in World War I

German scouts, from a postcard produced during World War I

way or road for the sole use of cyclists and other messengers. Also, when messenger cyclists have to travel with infantry columns, Major Thiess advised that they push their machines, as "cycling at the [slower] pace of the infantry produces a certain strain, in the long run."

Thiess was impressed by the guerrilla warfare carried out by bicyclists, especially the highly effective combined use of bicycle troops and engineers to raid railroads, bridges, trains, and airfields. He related numerous instances of German troops encountering advance patrols of Belgian cycle troops in the early weeks of the war, and found the effectiveness of French cyclists "extremely instructive." He described several incidents in detail, including a rapid pursuit action on September 14, 1914, by French cyclists crossing the Marne. At one point they covered some 87 miles in 24 hours, conducting surprise attacks on German ammunition carriers.

As to German exploits, their cyclists spearheaded a September 6th, 1914, attack toward the Seine, leading the cavalry, riflemen and engineers. Several times German cyclists were in the forefront of troops arriving at bridges, and prevented them from being destroyed by the enemy, such as at Ribecourt and Chateau-Thierry. As Thiess summarized it, when there was a race to a bridge or toward the sea, "again and again we find the cyclists hurled against the enemy." The Major also noted the ability of relatively small groups of cyclists to defend large sectors, citing the Germans at the Battle of the Aisne in 1914, and the Belgians at Haelen. The German cyclists were moved rapidly from sector to sector along a front to simulate a much larger infantry force spread along the line.

The relative stalemate of trench warfare did not lower the importance of cyclist units, in Thiess' view. Indeed, many German cyclist units subsequently assumed the role of rapidly moving reserves. In particular, Thiess described a "most famous" instance of the mixed use of cyclists and cavalry reserves at Limonava, on December 11-12, 1914. When a gap in the German line was threatened by an entire enemy corps, nearby German cavalrymen were thrown against them while a distant bicycle battalion was alerted. The cyclists rapidly entrained to a nearby point, disembarked, and pedaled to battle. The enemy was surprised by the numbers it faced, and retreated. Thiess concluded that the French, who absorbed many of their cyclists into other units as the war dragged on, "committed a great error which they had to acknowledge later."

In *Die Radfahrtruppe* Major Thiess laid down a few key rules

An Austro-Hungarian cyclist machine-gun section in Galicia

Germans on the march in Poland, World War I

for cycle units. He emphasized the importance of cycle troops approaching as closely as possible to the enemy before dismounting and fighting, but not so close as to expose the machines to undue fire, or have them overrun and captured. If resistance is light, bicycles can be taken virtually to the firing line. However, the machines should rarely be taken into combat off road. When bicycles have to be laid aside to fight, they should be camouflaged if at all feasible. Once combat is broken off the men must return to the bicycles as soon as possible. When cycle troops are advancing against the enemy, on foot, reserve soldiers should push the machines forward, rather than have the cycle troops return for them. If there is no one to help advance the machines toward the fighting troops then the entire unit should return together—not in fragmented groups—to get them. The act of returning to and mounting the bicycles was considered a great "moment of weakness." Soldiers should do so only when enemy action against them is not possible, and a party should always cover them. He also emphasized the value in using trucks to move bicycles and bicyclists rapidly from one area to another.

Major Theiss pointedly noted the cyclists' advantage in carrying out rapid, encircling movements of enemy forces, and

Military cyclists at the Russian Front, carrying folding bicycles, World War I

listed flanking maneuvers as "a typical use." World War I experiences also clearly demonstrated that when an enemy cavalry and/or infantry were retreating, cyclists could keep up a continuing pressure and create panic among the slower moving forces. Conversely, in providing a rearguard defense for their own troops during retreats, Thiess felt cyclists were particularly effective. They could delay the enemy much longer than other rear guard groups, since they were able to disengage from the enemy and return to their main forces more rapidly on their bicycles than could cavalry or foot soldiers. However, infantry are relatively flexible as to the direction of retreat, whereas cyclists must return to their machines if they are to continue operating as a cycle unit. He specifically cited instances in which Italians at Meata in 1916 and French troops at Roye in 1918 lost their bicycles and mobility because the combat situation forced them to retreat elsewhere.

Late in the war cyclists were used on both sides in conjunction with tanks and armored vehicles. In 1918, at Nieppe, the Germans used armored cars fitted with machine guns to carry out a thrust against the French, and the Italians combined bicycles with tanks in early November 1918. Although there was only limited experience to draw from, Thiess concluded—as had H. G. Wells some two decades earlier—that cyclists could and should be used with tanks, under certain conditions. Indeed, he believed that tanks were the most suitable escort arm for large bicycle units. In the 1930s the Germans adopted several of Thiess' suggestions and much of his philosophy. It was unofficial initially, since the World War I treaty forbade the Germans from building tanks. Before they eventually began to build them anyway, they tested their theories on tank tactics by constructing lightweight metal box frames made from bicycle tubing, with bicycle wheels on each corner. They covered them with wood and cardboard to simulate tanks, and had men inside push them around during field maneuvers.[9]

Germany's 1936 occupation of the Rhineland was an essentially domestic precursor to its subsequent foreign expansion. After the First World War all of West Rhineland, and a 30-mile

German troops on the Eastern Front, World War I

zone to the east of the Rhine River, was declared demilitarized. An allied force patrolled the region, in which no German forces were to be maintained, and all fortifications were taken apart. By the mid-1930s the allied forces had left. Ten months later Hitler re-occupied the Rhine. Despite outcries from many diplomats, there was little anyone could, or would, do about it. The British justified it by saying that the Germans were only returning home, not invading foreign territory. On Saturday, March 7, 1936, Cologne's Cathedral Square echoed to the sound and spectacle of goose-stepping troops. Few, however, remember the silently pedaling soldiers that preceded them through the square.[10]

World War II: Mobility and Bicycles

During a banquet toast Joseph Stalin emphasized to Winston Churchill that World War II would be one of movement, "a war of engines and octanes." The average American Division would employ 187,000 horsepower. For Germany, which in the late 1930s still got about 90 per cent of its energy from coal, the implications were obvious. It intensified efforts to develop synthetic fuel from that coal and by 1940 was producing some 72,000 barrels per day, or 46 per cent of its total oil supply. (Crucially, it accounted for 95 per cent of its aviation gasoline). Germany's eventual invasion of Russia occurred in part because it saw Russia as a threat to the Ploesti oilfields in Rumania, a German ally. Understandably, during their eastward thrust the Germans made Russia's Caucasus oilfields a prime objective.[11]

The Germans built their offensive around air support, fast tanks, and rapid infantry movement, and their achievements were astonishing. They took Poland in 27 days in 1939. Then, beginning in spring, 1940, they conquered western Europe in less than three months. They overran Denmark in one day, Norway in 23, Holland in five, Belgium in 18, France in 39, Yugoslavia in 12, and Greece in 21. The juggernaut was finally stopped only in the seemingly interminable, frigid spaces of the Soviet Union and on the hot, arid North African desert.

However, many have argued that the German Army's successes did not in fact reflect a superior military force, so much as an unprepared opposition. After the successful strike upon Poland, in 1939, General von Brauchitsch pointed out to Hitler that his army was in a bad state of affairs in terms of morale, discipline, equipment and supplies. It had expanded too rapidly, and any thoughts of a winter attack on France and Britain would have to be delayed. When Hitler did attack in spring, 1940, Britain and France in fact had more tanks than the Germans, and were outproducing Germany in aircraft and ammunition as well. But the Germans used what they had to much better effect, with superior strategy and tactics. In the long run, however, they did not have "the essential sinews of industrial war" to win. The great period of military expansion from 1933 to 1939 really needed another four years to be converted from merely large numbers to a truly superior "qualitative basis for military endeavour."[12]

Despite the reputation of the highly motorized blitzkrieg tactics, the German Army actually suffered from an acute shortage of motor transport. Only a handful of its infantry divisions were fully motorized, and even the best equipped of those supplemented that transport with nearly 5,000 horses. Overall, Germany employed some 2.7 million horses in World War II, double

This quiet, portable, two-man electricity generator could readily be moved about to power small dynamos for signalling, etc.

German soldiers, 1920s

the number it used in the Great War. Divisional artillery units and much of the supply service depended almost totally upon the animals to move weapons, ammunitions, and goods. During the invasion of Russia, German artillery and supply units harnessed up some 750,000 horses—and ate 100,000 of them over the winter of 1941-1942. Hence, the rapid and spectacular panzer strikes were supported by a relatively slow moving backup which consolidated the gains. Most German soldiers were dependent upon trains, their feet, or bicycles for getting around. If it was any consolation to them, they were not alone. On the other side, Russia had 30 divisions of horsemen, totaling some 210,000, and used ponies to draw supply sleds.[13]

At the start of the war the German army used large numbers of bicycles in various units. Partially motorized bicycle battalions were added to each cavalry regiment, with a motorcycle section transporting the heavy machine guns and 50-mm light mortars. Each infantry division's reconnaissance battalion included a bicycle troop. The mountain divisions, which ranged from using mules in higher and rougher country to motor vehicles in flatter areas, also relied heavily upon bicycles for reconnaissance; each mountain battalion had two or three bicycle companies. At

German bicycle-mounted gas detection unit

the General Headquarters level, the mobile battalion normally included two bicycle troops, and by 1939 there were 15 cavalry regiments, each with a bicycle battalion. As well, four man gas detection squads (*Gasspurtrupps*) with light protective clothing, gas detectors, and gas warning devices were found at all divisional levels and in all units. Many were bicycle mounted.

During combat the infantry tried to stay as close as possible to the tanks so as to take advantage of their firepower and

Germans negotiating a river crossing in France, 1940

paralyzing effect upon the enemy. However, the tanks did not slow their average advance so that infantrymen could keep up, but instead tended to advance in spurts and halt intermittently for the infantry to catch up. The value of bicycle troops in such situations is clear when the German Army's calculation of troop travel rates is reviewed. It showed that infantrymen averaged three miles per hour and typically covered about 20 miles a day, whereas cyclists pedaled along at an average of eight miles per hour, or double the infantry rate. That was a significant difference, given that tanks moved at an average rate of 12 miles per hour. The slowest links were the animal drawn support columns. While horse-mounted troops could cover some six miles per hour, draft horses could manage only about 12 to 15 miles per day per one-team wagon, with oxen even slower.[14]

Advancing infantry divisions stressed the importance of

Russian cycle troops

reconnaissance and establishing a trunk telephone line along the principal line of march. The reconnaissance battalions reconnoitered ahead of the division's front and flanks, using radio communications and messengers on horses, bicycles and motorcycles to keep in touch with the main forces. Horse mounted patrols usually operated cross country, while bicyclists used the roads. Reconnaissance battalions generally functioned well, although the combination of horse, bicycle, and motor vehicle units, with their varying rates of travel, sometimes proved difficult to coordinate.

The German blitzkrieg of 1940 forced some 220,000 retreating English troops on to the beaches of Dunkirk, along with half again as many French and Belgians. When the allies were evacuated they abandoned innumerable cars, trucks and bicycles. Afterward, the English *Army Quarterly* published an editorial urging the authorities to formally take up the bicycle for domestic defense, since the locals now faced the "imminent danger of the appearance of hostile raiding parties" from across the English Channel. Men up to 65 years of age were being recruited for

On the Russian Front, October 1941

home defense efforts, many unable to run or march long distances. Bicycles would not only enable them to move about more easily but would be much more viable than motor vehicles should roads and bridges be damaged.[15]

Earlier that year the Finns had been successfully resisting the Russian invasion of their country for two and half months. As spring approached the Finnish Army began collecting bicycles to replace skis during the anticipated summer warfare. On February 18, the Civil Guard appealed to civilians to contribute every machine that they could spare. Finland's relatively flat landscape, with "few main roads, but many paths through fields and forests", was readily amenable to cycle use, and it was hoped that Finland's soldiers would be as successful on them as they had been on skis. The Finns never got to find out; their army was forced to surrender to the Russians three weeks later.[16]

The following year, in Russia, the Germans launched the last of the blitzkriegs. However, with the exception of the Warsaw-Moscow highway, there were virtually no sealed roads, and the onset of the muddy period, in autumn 1941, bogged down

German cycle troops on a road in Holland, 1944

much of their army. Notwithstanding the atrocious surface conditions, which forced the Germans to increase the number of horses in use in each infantry division, bicycles still played a role. As Harrison Salisbury noted, a large number of the German soldiers who placed Leningrad under siege had pedaled there.[17]

By 1944, the war was going very badly for the Germans. A forced reorganization resulted in greater reliance upon horses and bicycles in infantry regiments and battalions. As well, units which previously had few or no bicycles—such as artillery regiments, medium and light artillery battalions, and antitank, infantry howitzer, and heavy weapons companies—employed them in increasing numbers. There was also the notable addition of bicycle heavy weapons companies; the 211 men moved themselves and their 75-mm howitzers, 81-mm mortars, and heavy and light machine guns on 29 horse drawn vehicles, 81 horses, and 181 bikes.

The most notable change in cycle usage was the creation of the Volks Grenadier Divisions, or People's Infantry Divisions, in 1944. They emphasized the "people's role," and stressed the emergency facing the Fatherland. They were smaller than previ-

German cyclists pushing bicycles across frozen ground on the Eastern Front

ous divisions, and had more small automatic weapons. The official divisional complement of 10,072 troops was equipped with 426 motor vehicles, 119 motorcycles, 3,002 horses, 1,142 horse drawn vehicles, and 1,522 bicycles. Specific bicycle allocations included 103 machines for the supply regiment, 100 for the 1,854 men in

each of the two infantry regiments, and 698 cycles for the bicycle infantry regiment's 1,911 men. There were 297 bikes for the Engineer Battalion's 442 men, and individual bicycle infantry companies were allocated 130 machines per 128 men. The former reconnaissance battalion was replaced by the Fusilier battalion, a

highly mobile, cycle-oriented unit, with 166 assigned cycles for its 200 officers and men.

In terms of the day to day use of bicycles, the Germans reflected a fact common to many armies: the higher up the command chain, the smaller the per centage and number of cycles in use. For example, a German regimental headquarters was allocated 38 bicycles, while a divisional headquarters had only five. Moreover, photographs suggest that the commanders of bicycle units were commonly driven about in motorcycle sidecars or other vehicles. In their defense, ever since the advent of the machines observers had noted the difficulty of commanding large number of rapidly moving cyclists en route, and motor vehicles for commanders certainly helped alleviate that situation.. On the other hand, it is much easier to be driven about than to have to pedal about. As Englishman observed in the 1930s in his own ranks, senior officers ride on army bicycles "comparatively little." Promotion has its perks.[18]

Facing page: German cycle troops negotiating a very rough mud and log surface

Normandy and the Aftermath

Prior to the Normandy invasion, the BBC broadcast a series of codes to French Resistance leaders. Among their ranks was Bayeux cycle shop owner George Mercader, who lived near the stretch of coast that would soon become famous as Omaha Beach. As a noted French racing cyclist, the Germans made an exception to the nighttime cycling curfew for Mercader, and allowed him to practice at anytime. On the night of June 5, at 10:15 PM, he heard on BBC radio that "It is hot in Suez," the signal to begin sabotaging telephone cables. He mounted his bicycle and pedaled quietly into the night to notify his colleagues to start cutting the local telecommunications links.[19]

Photographs taken before, during and after the Normandy landings illustrate the widespread military use of the bicycle by the Allies—from cycle mounted officers watching embarkation procedures, to soldiers inspecting bombed out German fortifications. There is no evidence that the Americans took any bicycles with them as part of the invasion fleet. However, there had been experiments with them years earlier. On October 10, 1941, an 88th Airborne Infantry Battalion was activated at Fort Benning, Georgia, to conduct, among other things, a series of tests relating to airborne cycle troops. Whatever the results, records do not show that the battalion ever served overseas or in combat, nor do records of later parachute/airborne infantry battalion tables of organization and equipment indicate that any of them had bicycles.[20]

In contrast, bicycles were prominent among the British invasion force and were used by both infantrymen and Royal Marine Commandos. The first official British photograph of the

British Army folding military bicycle

landings in France—and seven of the first eight—feature British Commandos jumping into the water with their bicycles and dragging them onto the beach. In his book on the D-Day invasion Stephen Ambrose refers to the bicycle's use and summarized the British machines as "cheap collapsible bicycles with baskets on the fronts to carry their rucksacks. The bikes had no mudguards, no pedals, just stems, and [Commando Peter] Masters found them damnable." The front racks were not actually baskets, however, but folding L-shaped frames onto which could be tied a variety of items. There was no mention of why the pedals were removed from the stems. Identical collapsible machines now held in the National Army Museum, for example, have their pedals.[21]

The 21,400 Canadian troops bound for Juno Beach on June 6, 1944 were also jammed in amongst bicycles. Many of the machines stood side by side, while others lay, several deep, about the floors of the landing craft. They included both folding and non-folding models. Interestingly, photographs show that most of the folding bicycles coming off the invasion craft, whether Canadian or British, were locked in the open position and rolled down the disembarkation ramps, rather than being carried off in the more compact, folded mode. Notes included with some Imperial War Museum photographs suggest the reason. Apparently the

Allied troops landing with bicycles at Normandy, June 6, 1944

Normandy landing, June 6, 1944

soldiers were informed that the landing craft would be approaching so close to the shore that it would be a "dry docking." Consequently the men had them locked in he open position, ready to be pushed down the ramps and onto the beach, for quick mounting. When they realized that in fact they would have to wade through quite deep water for some distance, there was another good reason that they did not take the time to fold them up and put them on their backs. As one soldier noted, there was heavy enemy fire, they had high octane petrol aboard, and three of the eight wooden landing craft around him had been hit and were burning. In those circumstances the soldiers pushed the machines off as fast as possible. In some cases the troops, heavily laden with packs and weapons, had to maneuver down extremely steep ramps into the water while trying to keep their footing and balance themselves and their bicycles. It proved very difficult for some to get their bicycles and gear through the water and across the sand.[22]

Once ashore, the French hedgerow, or *bocage*, country,

posed unexpected problems. On the Cotentin Peninsula the hedgerows created extremely dense, high barriers between the roads and adjacent cultivated fields. The fields averaged about 1.3 acres each, with some 500 of them per square mile. The roads between them were rutted to as much as 20 or 30 feet in depth. Tanks and other mechanized vehicles had a hard time breaking out of the roads, and proved vulnerable to anti-tank rockets. Only with the development of tanks fitted with special blades, much like a bulldozer, could they quickly break into the surrounding countryside. As well, it rained heavily through much of June, and the first half of July, leaving roads extremely muddy. In effect, the allies had to fight their way across relatively small pockets, one after the other, with great difficulty. It was not ideal for bicycling.

Nonetheless, as the invasion force drove the Germans southward and eastward, bicycles were widely used on both sides. British Commando Peter Masters recounted "pushing my bicycle and running over everybody who happened to be in my way" as he attempted to make it to the cover of some woods while under fire. In contrast, he later appreciated the pleasure of pedaling along a paved road without threatening fire. Masters noted that "a number of the commandos were riding German bicycles, army issue, heavy black things, much better than ours; their rightful owners had abandoned them galore by the side of the road. Some of our chaps were mounted on colorful civilian bikes, ladies' bikes, anything that would do to get to Varaville."[23]

As allied air support intensified, the Germans increasingly relied upon bicycles to help move soldiers around, in an effort to counteract the gas shortages and increasing destruction of the rail network. As the allies pressed onward, the roads became clogged. The accumulating numbers of dead horses, horse drawn carts, and destroyed motor vehicles sometimes halted even half-tracks and tank destroyers until the routes could be cleared by bulldozers. In contrast to the heavily motorized Allied advance, large numbers of retreating Germans took to horses and bicycles in their efforts to escape. However, at one stage the British covered 240 miles in six days during their re-taking of Belgium. At that rate of travel even horses were useless for retreating Germans. Journalists commented upon the numbers of German soldiers found

Allied soldiers inspecting a captured bicycle with anti-tank ammunition strapped to it, April 1945

dead alongside their bicycles, even as French civilians grabbed them to flee the fighting. As the First Army's operations officer wrote, "There was a quality of madness about the whole debacle of Germany's forces" in that summer of 1944.[24]

In late August General Patton's rush to the Rhine was halted in its tracks for several days when Eisenhower chose to give the available gasoline supplies to the United States First Army, which was supporting General Montgomery. Basil Liddell Hart concluded that "The best chance of a quick finish was probably lost" as a result. Patton's tanks were 100 miles nearer to the Rhine than the British, none of the bridges had yet been prepared for demolition, and the Germans were in a rout. They subsequently had time to regroup, and resistance stiffened. The war took another nine months to wrap up on the Western Front, and some three-quarters of a million allied casualties were suffered. Many believe that Eisenhower made a poor decision. He may have come to that conclusion himself, in light of Montgomery's lack of speed on the left flank. At a meeting with Churchill, General Eisenhower asked him to "persuade Monty to get on his bicycle and start moving."[25]

In the end, many of Germany's troops attempted to leave as they arrived—on bicycles. For some, the stay had been short. William Van Hoy, an American soldier, found himself crouching

German mountain troops evacuating Oslo, Norway, May, 1945

These German youths, left behind to ambush allied tanks near Weser, were captured on April 7, 1945

in a bunker with two Germans he had just captured, while American artillery fire continued to rain down upon them. The boys were only 17, and had just pedaled for three weeks to get there from Germany. "You know, I actually felt sorry for them," said Van Hoy. When it was all over, the allies confiscated thousands upon thousands of two wheelers either abandoned or taken from captured German troops, and gave them to local civilians. On the Russian front, the story was the same. Once-proud SS-Cavalry brigade soldiers retreating before the Russians used bicycles. In the more difficult road and surface conditions there, however, they spent much more of their time pushing.[26]

6. When Tojo Came A-Wheeling

"Even the long-legged Englishmen could not escape our troops on bicycles."
      COLONEL MASANOBU TSUJI, SINGAPORE: THE JAPANESE VERSION

The safety bicycle's most effective and widely publicized use by infantry took place over half a century after its development, in hot, tropical Malaya. There, after coming ashore in early December 1941, Japanese soldiers picked up the ubiquitous machines from local towns, villages and rubber plantation settlements and rode them south to Singapore. Along the way, they pedaled into military legend, stunning their British Empire enemies with the speed of their advance and ability to carry out rapid flanking maneuvers.

In November 1939, Winston Churchill had emphasized to his Cabinet colleagues that Singapore was a fortress that could only be taken by a siege of 50,000 men. Such an effort was adjudged forlorn and could be terminated at will by what were widely assumed to be superior British naval forces. They did not consider it possible that the Japanese would embark on such a mad enterprise.[1]

They did. One and a half hours before the attack upon Pearl Harbor, Japanese troops landed in northeast Malaya. Two days later the illusion of British naval supremacy in the Far East was shattered when the battleships *Prince of Wales* and *Repulse* were sunk. Sixty eight days later Singapore fell to a force of only 35,000 Japanese soldiers. It was the most intensive and significant use of cycle-mounted troops in military history. The bicycle, given little consideration by British military planners since World War I, gave the Japanese an immense advantage in tactical mobility on the Malayan Peninsula. It was, as one Australian writer observed, the time "when Tojo came a-wheeling." Many simply could not believe it. Others could not believe it could be so simple.

Japanese cyclist unit (possibly in China)

The Bicycle in Japan

The Japanese experience with the bicycle—using it, making it, and selling it abroad—has a long and controversial history. During the world-wide cycle craze of the 1890s, several English and American manufacturers established agencies in the country to compete with local "manufacturers" who were assembling machines from imported parts. Some postmen and policemen had taken to riding them, and the Japanese army was reportedly experimenting with them.[3]

By the end of the Russo-Japanese War a substantial Japanese bicycle industry was in place. In 1907 35,000 complete bikes were imported into Japan, along with enough parts to assemble another 25,000. The standardization of parts in 1910 led to the establishment of many small cycle factories using relatively unskilled labor and simple machines and tools for manufacture and assembly.

In 1936 Teijiro Uyeda and Hiroshi Koyasu prepared a report on the Japanese cycle industry. They indicated that the 1913 production level increased twenty seven-fold through 1918. By 1935 some 37,000 workers had doubled the production yet again to nearly 25,000,000 yen annually—and that did not include factories of less than five workers, which Uyeda and Koyasu estimated accounted for another 10,000,000 yen's worth of cycle output. Bicycle exports accounted for over half of the production. A specific number of machines was not cited, and extrapolation from Census of Manufactures data is difficult (it combines parts and completed cycles, in various combinations). However, several hundred thousand bicycles a year were being exported, in particular to India, China, and Southeast Asia.[4]

Japan was accused of economically "dumping" its machines, and engaging in deceptive practices, in order to penetrate overseas markets. In England, in 1934, an exhibit of Japanese goods included an imitative cycle bell "bearing what appeared to be the emblem of a well known Birmingham firm." It was also announced that in a few days an "experimental consignment" of 500 Japanese bicycles would be landed, with 5,000 more possibly following; they would be considerably less expensive than English-made machines. The National Association of Cycle Manufacturers considered it a hostile act, and instructed its members not to sell or repair Japanese cycles, tires, or accessories. They felt that cheap Japanese imports were "the real yellow peril of the present day."[5]

The British views reflected those in America. In June, 1928, the U.S. Federal Trade Commission issued a complaint against the K & K Supply company of New York for unfair competition—it assembled bicycles from a combination of unmarked Japanese and American parts. There were often no name plates on the cheaper, imported Japanese frames (to which U. S. manufacturers's plates could be attached by the unscrupulous), and the required marking, "Made in Japan," was stamped on the heads of the frames in such a way as to be easily covered or removed. That was not acceptable in the self-avowed land of free enterprise.[6]

The fears of the American and British manufactures and retailers were probably well founded. The potential effect of cheap Japanese bicycles was clearly demonstrated in the Australian market in the 1930s. For example, in Collie, Western Australia, local miners used bicycles in large numbers to commute to a colliery a few miles out of town. The community's major cycle dealer reported that the less expensive Japanese machines quickly captured 25 percent of the market.[7]

The Malayan Campaign

Reports that the Japanese were using bicycles in their Malayan invasion were not initially believed by some. Many found the fact astounding. Others eventually grasped the significance. As Major-General Kirby summarized it, "the Japanese speed of movement, their ability to overcome obstacles ... came as a complete surprise." Whatever the initial perceptions, the bicycle's employment was ultimately to confound the British forces. And, somehow, it seemed not quite right. Winston Churchill, in *The Hinge of Fate*, conveys an underlying indignation that the mundane tool could have played such a role in the fall of the Far Eastern Bastion.[8]

However much Churchill might have been irritated by the Japanese resort to the device, his own experience certainly left him aware of the machine's potential in such circumstances. At the conclusion of a journey through East Africa, Churchill wrote in 1908 that:

> the best of all methods of progression in Central Africa—however astonishing it may seem—is the bicycle ... From my own experience I should suppose that with a bicycle twenty-five to thirty miles a day could regularly be covered in Uganda, and, if only the porters could keep up, all journeys could be nearly trebled ... In the dry season the paths throughout the bush, smoothed by the feet of natives, afford an excellent surface ... the track is only two feet wide, and ... the bicycle skims along ... at a fine pace; and although at every few hundred yards sharp rocks, loose stones, a water-course, or a steep hill compel dismounting, a good seven miles an hour can usually be maintained.[9]

Japanese on the Malay Peninsula

Japanese unit entering Batavia, 1942

In Malaya the Japanese did not wait for porters, and in lieu of bush paths they used the extensive network of very good British-built highways and agricultural estate roads.

Unlike Churchill, who only discovered the merits of East African bush cycling after arrival, Colonel Masanobu Tsuji, chief architect of the Japanese invasion, had done his Malayan homework carefully. The fact was later noted by Churchill, in 1951, when he wrote that there "had been minute pre-war study of the ground and conditions. Careful large-scale plans and secret infiltration of agents, including even hidden reserves of bicycles for Japanese cyclists, had been made."[10]

In his book, *Singapore: The Japanese Version*, Tsuji denied that he had hidden bicycles in the region. His dismissal of Churchill's claim rings true. In fact, the machines were commonplace and spare parts and replacement cycles were readily available throughout Malaya. Bicycles were there for the taking, and his

Japanese cycle trailer, cyclists and infantry making a river crossing alongside a collapsed bridge in Burma

Troops on a road in the Philippines

troops simply took them. The Japanese Defense ministry's history of the Malayan campaign specifically noted that invasion troops were refused permission to take bicycles on the ships, even when wanted for scouting immediately upon landing. In a personal memoir, Yoshiki Saito, a Japanese soldier, said that no bicycles were allowed on his ship, as they were known to be plentiful in Malaya. He and his fellow soldiers quickly converted themselves to a bicycle squad by collecting machines from a village, helped by a local Japanese resident.[11]

In F. Spencer Chapman's *The Jungle is Neutral*, a British soldier who remained on the peninsula throughout the war as a guerrilla fighter, described Japanese soldiers 'systematically searching the road-side kampongs, [rubber] estate buildings, and factories for bicycles." Lieutenant Colonel P. W. Thompson put it more succinctly: "In the Japanese advance a Malayan bicycle seen (and there were many to be seen) was a bicycle commandeered." And a Royal Australian Air Force officer noted, in his diary, the fruitless effort of the British to stem the Japanese use of bicycles by buying them from the locals and destroying them: "High rates are paid to encourage this destruction, since Japanese infantry and others have got through on cycles dressed as Malays."[12]

Tsuji devoted a chapter to describing the bicycle's role in the invasion of Singapore. In lieu of the horse transport many of his troops had used in China, he reorganized each regiment so that all heavy materials could be loaded on about 50 trucks. All officers and men not riding the trucks used cycles. Ultimately each division ended up with some 500 motor vehicles and 6,000 bicycles. Each cyclist carried upwards of 65 pounds on his machine, as well as a light machine-gun or rifle. Chapman noted that their weapons were normally tied to the frames of the bicycles, so that they took some time to go into action when suddenly attacked. He described their advance as consisting of "waves of cyclists and motor transport." The majority rode in groups of 40 or 50, three or four abreast, "talking and laughing just as if they were going to a football match." They traveled lightly and were equipped with a motley assortment of gear, but gave an impression of "extraordinary determination." The only standard equipment he observed was a mackintosh cape with hood that covered the rider and paraphernalia. They pedaled right through the tropical downpours.[13]

Tsuji's essential tactic was to rely not upon traditional artillery bombardments preceding the troops, but upon lightly armed forward troops, moving quickly ahead of their tank support. The bicycles were integral to the tactics. The ability of the cycle mounted troops to outdistance tanks and other support vehicles is reported in numerous accounts. As a western journalist noted in one incident, "by the week-end infiltrations of bicycle patrols on the east coast had been reinforced by heavier metal." The Malayan Campaign, in short, "was an excellent example of the proper employment of well-trained, aggressive, and highly mobile troops, led with bold imagination and ably supported."[14]

As well as landing on the northeast Malayan coast, Japanese soldiers disembarked the same day at several locations in Thailand, and headed towards the west coast of Malaya. This was the principal thrust, which utilized the good road and rail networks there, and was shielded from much of the force of the northeast monsoon rains by the central mountain spine of Malaya. The cycle columns rolling southward down the west coast sometimes got into trouble because of their rapidity of movement—a fact well illustrated in an ambush set up by Australians at a crossing of the Gemencheh River, on January 14, 1942.

A company of the 2/30th Battalion went three miles up

the road beyond the bridge, toward the Japanese advance, and hid. In the late afternoon some 300 Japanese cyclists rode past the ambush position; after an interval another 800 or so followed. The "blithely chattering Japanese push cyclists, riding five or six abreast ... resembled a picnic party rather than part of an advancing army, except that they carried arms." The Australians then blew the bridge and sprang the trap on the completely surprised Japanese, whose "rifles and automatic guns were strapped to their cycles." The invading forced suffered nearly 1,000 casualties. The Australian ambush, one of the few successful momentary hindrances to the Japanese advance, was eventually broken up by advancing support tanks.[15]

The cycle troops had their problems. Tsuji reported that heat contributed to numerous leaks and punctures, and each company included a bicycle repair squad of at least two men, who averaged repairs to about 20 machines a day. When in a hurry, flat tires were often discarded and the troops rode on the rims. On hard dirt and sealed surfaces it was quite practical, and several accounts mentioned that the noise of several bare-rimmed bicycles on bitumen sounded, at a distance, remarkably like that of advancing tanks.

A particular advantage lay in the relative ease and rapidity with which cyclist troops could overcome various obstacles. At blown bridgeheads, temporarily impassable to motor vehicles, troops could wade bicycles across shallow streams, carry them over lightweight makeshift bridges (some no more than logs supported on the shoulders of colleagues), or ferry them across in small boats. In that way continual pressure was kept upon retreating British forces. When soldiers dismounted to fight, guards were left with the bicycles. As the Japanese fighters pressed onward, the guards would obtain the "co-operation" of local residents to carry, push or ride the machines forward.[16]

Ultimately, the well developed network of rubber plantation roads and narrow paths made it possible for Japanese cyclists to quickly outflank allied troops on the main highways. Raymond Callahan's analysis of the fall of Singapore laid great stress upon that fact; British forces, "lavishly equipped with mechanical transport ... were road-bound," while the Japanese cyclists cut off re-

Japanese troops cycling into Singapore, 1942

treating British and allied forces along the main roads by quick encircling maneuvers through the rubber estates. In trying to stop the advance, the British "pinned our faith to the few roads but the enemy uses the tracks and paths, and gets around to our rear very much as he likes." In essence, they had superiority in tactical mobility. But it was not always easy for the Japanese. As Chapman described it, travel through the rubber estates was often "along tiny muddy paths which shot suddenly downhill to cross irrigation ditches or spanned them perilously on narrow plank bridges."[17]

Tsuji's summary of the situation was that "thanks to Britain's dear money spent on the excellent paved roads, and to the cheap Japanese bicycles, the assault on Malaya was easy." Those

Japanese soldiers killed and wounded might not have agreed, but it was clearly a perceptive move on Tsuji's part. The 600 mile advance was expected to take 100 days; the "blitzkrieg on bicycles" required only 70. The fall of Singapore was at once the greatest land victory in Japanese history and the "most catastrophic defeat in British military history."[18]

The Japanese success in the Malayan invasion, in John Keegan's words, "resulted from the flexibility and dynamism of their methods, akin to those that had characterized the German blitzkrieg in France in 1940." The British in Malaya did not help their own cause, making some bad decisions, having some poorly trained and officered Indian troops, and leaving airfields intact during retreat, among other things. And the Japanese had air superiority and 57 tanks, which the British did not. But the principal reasons for the defeat, in virtually every assessment, were the Japanese soldiers's speed, tenacity and mobility. Although outnumbered two to one, they grabbed local trucks, cars, and bicycles as they went, and used fishing boats along the coast as well, both to push forward, and to skirt around to the sides and rear of retreating British and allied forces. When considered collectively, most comments upon the nature and rapidity of the advance point to the bicycle as the crucial transport element. It enabled the Japanese infantry to keep up an unrelenting pace not possible on foot. Numerous on-the-spot accounts noted that the cyclists were routinely well ahead of their tank and motorized support. In contrast to the European proving ground, the Malayan blitzkrieg was spearheaded by bicycles, followed by tanks.[19]

Roy Dilley noted that "the speed and apparent ease with which they inflicted a series of defeats upon ... allied forces following the outbreak of war, created among their opponents a feeling that Japanese soldiers were invincible in jungle operations." Certainly that was the impression left by the Singapore invasion, during which a reporter for *The Bulletin* (Sydney) described them as "often six-footers." In reality, Japanese soldiers averaged only 5 feet 3 inches in height, weighed about 120 pounds, and few in Malaya had previous jungle fighting experience.[20]

George Aspinall, whose prisoner-of-war photos taken in Changi and on the Burma-Thailand railway are legendary historical records, wrote about his initial contact, in Singapore, with the conquerors of the Gibraltar of the East:

> I didn't actually see any Japanese until late in the afternoon of that first day. I'd driven my ambulance out of the St. Andrew's Cathedral grounds and parked it beside the road. I was standing at the front of it, leaning on the bonnet, when a group of Japanese soldiers rode by on bicycles ... that was my first close-up experience with the Japanese as a prisoner-of-war.[21]

Tsuji's adoption of bicycles was, John Toland believes, a typical, "shrewd, on the spot" decision made during his pre-war reconnaissance and personal assessment of the Malayan situation. Tsuji had had many years experience throughout Asia, where the bicycle was widely used domestically, and would have been alert to its intrinsic characteristics. Further, the Japanese army had experience with cycle units in China, and Saigon was occupied in part by two-wheeled Nipponese soldiers, facts rarely alluded to by historians. Tsuji was aware of such matters, and combined with his first-hand knowledge of the machine's capabilities in rural Asia, would have been attuned to its military potential.

Bald, bespectacled Masanobu Tsuji "made a commonplace of eccentricity," and veered from one side to another of that line that separates genius from madman. On the one hand he was susceptible to righteous, moral outrage; on the other, he carried out dangerous reconnaissance missions to gather information for his assessment of the military in Malaya, and consequent invasion plan. A dark side of Tsuji included a racial fanaticism aimed against the white colonialists and their Asian "allies." In a manual which he prepared, he advised his invading troops that they would see for themselves how money "squeezed from the blood of Asians maintains these small, white minorities in their luxurious mode of life . . . here is the man whose death will lighten your heart." After the capture of Singapore, Tsuji was responsible for the execution of thousands of Chinese. Later, in the Philippines, he convinced a number of Japanese officers to give illegal orders to kill captured Americans and their Filipino supporters. Not all Japanese soldiers respected him. Lieutenant General Sosaku Suzuki, Yamashita's Chief of Staff in Malaya, felt that Tsuji's "speech and conduct

Tsuji and Beyond

Masanobu Tsuji

Australian with an abandoned Japanese bicycle, New Guinea

were often insolent" and disliked'his inhumane treatment of the Chinese. Suzuki thought that as long as Tsuji and his like-minded colleagues exerted influence on the army, "it can only lead to ruin. Extermination of these poisonous insects should take precedence over all other problems."

Whatever his character failings, Tsuji's brilliance in putting the overall invasion plan together lay in thinking beyond traditional military patterns involving infantry, artillery, cavalry, and armored units. Tsuji's integration of bicycles into his advance has been considered by some to be a stroke of genius. On the other hand, it is possible that, for an army operating with constraints on gas supplies, he had little alternative. It may have been only a straightforward, pragmatic attempt to resolve a problem. The fact that the cycle-mounted troops succeeded, possibly beyond even his expectations, would make it look brilliant only after the fact. Regardless, the result was the most dramatic and effective use ever of the bicycle in wartime. One can not help but wonder how other military planners in Tsuji's shoes would have reacted.[22]

The Japanese use of the bicycle in Malaya is so widely

known that it has almost become a cliche in summarizing that campaign. But the machine was also widely employed by Japanese troops in other theaters of operation, as written accounts and photographic references attest. However, the limited evidence available from elsewhere does not clarify to what extent the bicycle was either a convenient afterthought, or an integral part of planning—nor how effective it proved to be. But whether adopted by accident

The Japanese use of the bicycle in Malaya is so widely known that it has almost become a cliché in summarizing that campaign. But the machine was also widely employed by Japanese troops in other theaters of operation, as written accounts and photographic references attest. However, the limited evidence available from elsewhere does not clarify to what extent the bicycle was either a convenient afterthought, or an integral part of planning—nor how effective it proved to be. But whether adopted by accident or design, in no other place did the bicycle play a role anywhere near as crucial or spectacular as in Malaya.[23]

In Burma, the mountainous terrain, heavy vegetation, and smaller network of good roads made the bicycle a less attractive proposition than in Malaya. However, it was still used to great effect. Photographs and accounts tell of large numbers of Japanese troops pushing heavily loaded bicycles, serving as messengers, and some towing light two-wheeled trailers behind their machines. An Australian writer in *The Bulletin*, commenting upon an advance in Burma in early February 1942, described the Japanese movement beyond a damaged bridgehead as "their usual dribble of manpower, which flowed on—on bicycle, on foot."[24]

In New Guinea, some 30 miles inland and 3,000 feet up from coastal Buna, at the foot of the ascent to Kokoda, "the Japanese were seen coming down the road to Awala. They wore green uniforms and steel helmets garnished with leaves. Some were on bicycles. Each carried in addition to his arms and ammunition a machete for cutting through the jungle, a mess tin of cooked rice, and a shovel slung on his back." They did not stop there. A photograph in the Australian War Memorial shows a bicycle nearly buried in mud along the notorious Kokoda Trail. It is a wonder anyone could even have considered using one in such circumstances, let alone have got it that far.[25]

Once the fighting had finished, Japanese bicycle patrols were routinely used in occupied territories during the remainder of the war. In northern Sumatra, for example, a group of commandos attempting to place charges on a bridge across the Peudada River had to wait anxiously for a train, Japanese staff car, and bicycle patrols to pass before successfully completing the job.[26]

The Japanese were not the only users of the bicycle in Asia during World War II. For over three years Colonel F. Spencer Chapman, of the British Army, remained in the Malayan jungle, working with other soldiers and Chinese guerrillas to harass the Japanese. His book, *The Jungle is Neutral*, is a fascinating compendium of the machine's use in wartime circumstances in difficult terrain, including along the mountainous spine of the peninsula. During major moves across country, the cycles were sometimes loaded with up to a hundred pounds of gear, in two army packs hung on either side of the rear wheels. On occasion the riders carried an emergency "getaway" haversack of essentials tied to the handlebars, to grab if they had to abandon the cycles and escape into the forest. Parts and tires were hard to come by, and routinely recycled. The result was collapsed forks, broken chains, and disintegrating wheels and tires.

Since the bicycle was widely employed by locals, Spencer Chapman found that by disguising his men as Malays they were able to move relatively freely, often pedaling right past Japanese guardposts. Aware of the machine's guerrilla role, the Japanese frequently forbade its use on the roads at night. Spencer Chapman relates how, in response, they once freewheeled down a long hill in darkness, hunched below the level of the handlebars to avoid being cut off their machines by wires strung across the roads by the Japanese.

The jungle may have been neutral, but the terrain, climate, insects and Japanese were not. As Spencer Chapman noted, "After our laborious days of wading through swamps, toiling up and down jungle ridges, and lurking in mosquito-ridden rubber estates, it was a marvellous sensation to be speeding through the night on bicycles."[27]

7. The Home Front: World War II

You can be the wealthiest woman in Paris and still have to depend on your bike, for there's literally no other way of getting around.
                    THE NEW YORK TIMES, SEPTEMBER 10, 1944

World War II, with its unprecedented and seemingly insatiable military demand for fuel, tires, vehicles and parts, left civilian public and private transport in dire straits on many home fronts, and in utter chaos in others. For a substantial percentage of the world's population, private motorized travel ceased for the duration of the war. In those circumstances the bicycle often assumed great importance. In many European countries cycling had long been an integral element of the local transport scene, and wartime domestic reliance upon it represented little change for part of the population. In countries such as America and Australia, however, the motor vehicle had substantially replaced the bicycle. There, the sudden, enforced reliance upon pedal power for personal travel was a radical shift for many. Ultimately, people everywhere faced the same fundamental issues with respect to bicycles as they did with virtually everything else: where to get them—if they could be had at all—and how to keep them working.

In Germany, the outbreak of war in 1939 signaled the onset of difficult times for the Third Reich's residents. Those fortunate enough to have motor cars had to convert them to synthetic fuel use. By late October, just seven weeks after the invasion of Poland, pressure upon Berlin's public transportation system was so great that authorities urged citizens to use bicycles whenever possible. They promised to build a network of cycle paths and provide special bicycle parking places in the city center, at industrial plants, and in the suburbs. Cynical bicyclists could have been forgiven their doubts; officials had proposed similar cycle lanes four years before, but few materialized.[1]

As the war ground on, private motor vehicle use virtually ceased, and bicycles became increasingly precious. Because

A victim of the Hamburg bombing and firestorm, July 1943

of the strict rationing of rubber and tires, authorities appealed to citizens to stop all non-essential cycling and give their machines to war workers. Members of the Hitler Youth Organization were asked to use theirs only for "business purposes." At least Germans could own cycles. In contrast, in 1941 the Berlin Chief of Police announced that Polish civilians and laborers living in the city could no longer "acquire, possess nor use bicycles." The conquered could walk.[2]

By 1944, as the allied bombing raids on German cities and towns intensified, people found that the bicycle was often the only transport device that could be used in the rubble-cluttered streets. As building after building collapsed from bombing and fire, it was possible to get down many roadways only by walking or alternately riding, carrying and dragging a bicycle over and around the debris. Motor vehicles simply could not negotiate the rubble-clogged streets, and there was no spare equipment to clear other than the major routes. In the end, much of Germany's urban surface transport network simply ceased to exist.[3]

France: The Resistance Role

As the Germans approached Paris, in June 1940, many Parisians panicked and sought any means of getting out of the city. Cycle shop owners were mobbed, with customers offering up to 2,000 francs for anything that rolled. The day after occupation they could be had for as little as 25 francs, and many lay abandoned outside train stations. Over the next four years the bicycle, already a staple of the French transport scene, became the very backbone of everyday life, and an integral element of the Resistance movement.

Within a month of occupation, the Parisian scene returned to its old routine, in some respects, as the air raid trenches along the Champs Elysees and in the parks were refilled. Curfew was extended to 11:00 PM during the long, warm summer nights, and the cafes were again busy, often filled with German soldiers. But on the streets, things were dismal. There was a dearth of taxis and motorized vehicles of any kind. Quite simply, gasoline had become "more precious than the rarest couture perfumes," wrote a *New York Times* correspondent. Those civilians still allowed to use motor cars frequently tied a bicycle on, in case the vehicle was requisitioned, or they ran out of fuel. At the railway stations a profitable new business developed overnight—wheelbarrow porters. They awaited the arrival of trains, and carted baggage to the travelers' quarters. As well, some 2,500 licensed velo-taxis (bicycles pulling rickshaw-like trailers) were brought into use. However, charges became extortionate, and authorities eventually stepped in. In July 1943, they reduced the number of licensed operators to

A Frenchman pulling a cycle cart, Bayeux, July 1944

800, standardized fares, and increased the general bicycle tax by 60 per cent, to raise additional revenue.⁴

The underground Resistance efforts appeared, at first, to be hopeless in the face of the powerful German forces. But the networks grew, the operatives became more skilled and organized, and their harassment and sabotage efforts were increasingly effective. By destroying or damaging such things as canal locks, railway switches, and factory machinery, the Resistance tied up German Troops and restricted the benefits they obtained from the French agricultural and industrial machine. Eventually the Germans had to deal with the same problem faced by the British in the Boer War, 40 years before: how to stop local resistance fighters and foreign agents who were using bicycles to carry themselves, messages, radio sets, and weapons about the countryside, with their innocuous bicycle pumps converted into single shot weapons, and small oil cans made into bombs.

Unlike in the Boer War, the bicycle was crucial to the

A bicycle generator used to silently recharge batteries for the Dutch Resistance, displayed after liberation

Parisians after liberation, 1944

domestic life of France. The millions of machines could not be confiscated if the economy were to continue operating. However, one immediate step the Germans took to control the bicycle's use by the Resistance was to require permits to ride them. They also decreed that no one could buy a bicycle without permission. Because that order would prove a hardship to many, the puppet French Vichy Government delayed its implementation until August 1941, over a year after occupation.[5]

Nine months later, in May 1942, the Germans finally forbade any use of bicycles after dark in over half of the occupied

Parisians cycling to the Arc de Triomphe after liberation, August 26, 1944

territory, in an effort to check increasingly violent French Resistance activity. The directive noted that those who had attacked German soldiers or committed sabotage usually rode bicycles during their escapes. The night cycling ban placed a harsh burden on city dwellers, however. Many of them cycled into rural areas to forage for fresh food, and without the machines their radius of travel during short winter days was severely restricted.[6]

The bicycle itself often determined the operating range of local Resistance agents. The historian, M. R. D. Foot, who has written extensively on the Resistance, referred to one landing strip used clandestinely by French and British operatives. By February 1944, all "safe houses within bicycling distance of the [landing] ground had been used and over-used to the limit of prudence," and it was decided to stop using the airstrip. A Frenchman, Dericourt, saw off the last flight on the night of February 3, 1944. Knowing that local network security could be compromised by the cycles left lying about by the departing agents, he remained behind "to help his two assistants dispose of thirteen bicycles before dawn."[7]

After the allied landings, in June 1944, the domestic situation became ever more grim as the fighting raged and the retreating Germans took whatever they needed or wanted. With the liberation of Paris, in August 1944, things improved rapidly. By September 1, a battalion of American Military Police arrived to help control traffic in the city. They were sorely needed. As gasoline began arriving, more civilian vehicles appeared on the streets daily, joining the American and British vehicles transporting supplies into the capital. The traffic problem was little short of horrendous. "The real problem," a correspondent observed, was "bicycles, thousands and thousands of them." Bicycle traffic at busy intersections "was a mess ... everyone doing just about what he wanted to." The Parisian cyclists, used to years of riding on almost automobile-free streets, were darting all over the streets and, in their enthusiasm, holding on to the military vehicles. Eventually gendarmes with whip batons were assigned to control "the waves" of pedalers.[8]

England: The Invasion Threat

When the English went to war in August 1939, German attacks on their shipping quickly took a toll on available goods, many of which were fully imported or supplemented from abroad. Authorities, however, were slow to impose controls. On November 30, 1939, *The Times* reported that rationing "is to begin gently, quite partially, and not immediately." It was another six weeks before the first items were restricted. The initial petrol ration was established at 1,800 miles per year. Eventually, though, no fuel at all could be used for non-essential private use, and what little petrol was available was increasingly inconvenient to get. To make it easy to destroy petrol supplies should the Germans invade, the British government closed 17,000 outlets, particularly in the southeast. To confuse invaders, all road signs were also taken down. It was all accompanied by a black market where, as one wrote, "anyone who had enough money and a small conscience could find where to buy petrol."[9]

As civilians came to rely increasingly upon bicycles, the British cycle industry, then the largest in the world, was flooded with both domestic and foreign orders. However, the English government began taking over many of the factories, the remaining cycle producers could not obtain sufficient materials, and the production of the machines eventually ceased altogether. At the same time, the shortage of petrol increased the demand for cycles. The newspapers were filled with classified ads seeking bicycles and parts. Offers ranged up to 15 pounds for cycles "of any make," "of any age," with "present operation not essential." The advertisements noticeably increased by mid-1942.[10]

One journalist, commenting upon the potential invasion, reported that German paratroopers landing in the low countries in 1940 had used folding bikes, strapped to their backs. As the Germans faced England across the narrow English channel, some felt that "so disastrous is the use that an invading force might make of bicycles," that the English populace was instructed on how to render their machines useless. Among the tips offered was that bicycle shops should remove all pedals from their machines, and individual cyclists should affix wing nuts to crucial parts, which could be easily removed and carried in the pocket when not riding. To drive home the lesson, a British War Department film showed a German soldier, disguised as a nun, suddenly whipping a folding bicycle from under the habit, assembling it, and pedaling off.[11]

A large number of Italian prisoners of war were interned in England, and many worked on local farms. The problem of how to get them to and from the farms each day was solved by giving bicycles to those traveling more than seven miles from their camp. They were sent on their way, unescorted. The War Office defended the practice—at a time when civilians were having trouble getting bicycles—as a way of "avoiding a considerable waste of gasoline and rubber" that trucking them would have entailed, and because there were no spare soldiers for guard duty.[12]

As the German threat of invasion receded, and air raids became less frequent, blackout restrictions were eased. From March 9, 1943, lights could be used on bicycles as long as the top half of the lamp glass was blacked. The decree was welcomed, and a publication was widely circulated showing how to properly mask a cycle lamp. Late in 1944, with the allied drive across the English Channel in full force, there was an urgent need for the supporting motor vehicles in England to be able to use their full headlights, running as they were on long schedules in the shortening winter days. Finally, on December 27, 1944, all light restrictions were removed. For night riding purposes, the war was over.[13]

Australia: The Manufacturing Effort

As in other countries, the Australian government established wartime controls on motor vehicles, petrol, and spare parts, and the population had to adjust accordingly. It was a jolt, for Australians were heavily dependent upon the motor vehicle in that vast, sparsely settled country. By 1930, the nation ranked only behind the United States and New Zealand in per capita car ownership. Petrol rationing began in October 1940, and within a year private car owners received only enough to drive about 15 miles per week. Some converted their motor cars to use charcoal or gas burners, and within two years there were over 12,000 such vehicles in the state of New South Wales alone. Others simply stopped using their cars. From 1939 to 1942, vehicle registrations dropped by 16 percent as the low petrol allowance made them uneconomical to license and maintain. And, as in England and America, some Australians resorted to hoarding or black market coupons to keep going.[14]

The public transport system was severely strained, and mandatory "queuing" was introduced to prevent passengers from rushing uncontrolled toward arriving trams, busses and trains. The government eventually reserved the busses for special classes of travelers, such as workers, and that put even more pressure upon the tram and train services. Film matinees finished by 4:30 P.M., so that moviegoers could leave the city before the workers. Taxi drivers, limited to 22 gallons a week, rarely took passengers anywhere where that they were unlikely to get a return fare. Thus, late evenings in Sydney often meant a walk home across the Harbour Bridge for northside residents.[15]

The basic transport mode for many was aptly summarized by an Adelaide headline: "Bicycles Have a New Importance and Popularity in War-Time." In Sydney, the National Roads and Motorists' Association undertook its own war conservation effort

Commuters in Melbourne, Australia

by putting two road service patrols on bicycles to save petrol. The mechanics, with large tool boxes fitted above a small front wheel, pedaled about the inner city area helping motorists.[16]

Within three months of the bombing of Pearl Harbor, a detailed report on the Australian bicycle industry was produced. Children's models were immediately banned, and all motorized

A bicycle-mounted road service patrol, Sydney

bicycles were channeled to the armed forces. Wartime cycle production essentially centered about Bruce Small's Malvern Star operations. Small traveled about Europe prior to the war, studied the military bicycles of the Germans, French and Italians, and attempted unsuccessfully to sell the Australian military forces on the

Military cycles bound for Allied Pacific forces

Australian newspaper advertisement

Malvern Star production facilities, Australia

idea of adopting them. He advocated specially trained cycle corps which would not be susceptible to the impending petrol shortages.[17]

In 1940, increased Australian involvement in the European war led to Malvern Star receiving its first order for military bicycles. In 1942, when General Douglas MacArthur took command of the Allied forces, in Melbourne, he immediately ordered an increase in bicycle production, much to Small's surprise. The Yanks were among the last he thought would want them. The Malvern Star cycle factories, the largest in the southern hemisphere, were stretched to the limit. As overseas supplies were cut off, they had to begin producing some components for the first time in Australia. The tubular manufacturing facilities at Malvern Star also received military orders for radio masts, ambulance stretcher carriages, tent frames and radio location sets (which needed complex tubular structures). Staff increased from 400 to 726, and Small and his executives donned overalls to work three night shifts a week in addition to their normal managerial functions. Military cycle production in Australia reached its peak in 1943, along with a soaring domestic demand. By the end of the war, Australians were bicycling on a scale never seen before or since on the nation's streets, and Malvern Star was proud of having kept "the home tires turning."

America: Rationing

America's home front was a great contrast to that of Australia and the European nations. As Edward R. Murrow noted, "We are the only nation in this war which has raised its standard of living since the war began." Its economic output was prodigious. The country was able to put 12,000,000 men into uniform, and equip and support them to fight in two separate wars, simultaneously. The navy that America built was bigger than the combined fleets of all its allies and enemies, and it supplied its allies with a high percentage of their wartime needs. Domestically, civilian spending increased by 20 per cent during the war. But spending patterns were altered. Production of many things ceased, others became harder to get, and some were rationed (civilians were limited to two pairs of shoes a year, for example). Perhaps the worst blow was when the Office of Price Administration (OPA) took possession of the 500,000 new cars left when the last car assembly line was closed, on February 1, 1942. For the remainder of the war it handed them out sparingly to those with special needs.[18]

After war with Japan broke out, the most mobile society on earth was reluctant to give up that mobility. It was three months before the use of gasoline for automobile racing was banned, and another two before general rationing was imposed on the east coast. That raised an outcry from the easterners, who felt penalized in comparison with midwesterners, where the greater distances made rationing less feasible. Tourist centers such as Miami and Atlantic City, which saw their economies threatened, complained loudly. And, until mandatory blackouts were finally imposed, Miami Beach kept its beachfront neon lights on for nearly half a year after the war started. Finally, in late 1942, a 35-mile-an-hour speed limit was imposed, and gasoline ration coupons were issued, allowing most motorists only a few gallons a week. Among the responses was the development of a black market, and counterfeit coupons.[19]

On March 12, 1942, three months after the attack on Pearl Harbor, the War Production Board (WPB) and OPA formally banned the manufacture of children's bicycles, but allowed a threefold increase in adult cycle output, to 756,000 annually. A bicycle was deemed to be "adult" if it measured more than 20 inches from the center of the pedal crank to the top of the saddle pillar post. By the end of that month, automobiles were being removed from the roads at an estimated rate of 17,000 per day, and cycle sales unexpectedly skyrocketed as more and more employees began pedaling to work, often to save their rationed gas for leisurely pursuits. Commercial parking garages offered protected bicycle parking, including routine inspection and servicing, for anywhere from $1.00 to $1.50 per month. Power companies put meter readers on cycles, and articles appeared in newspapers advising on proper care. Consequently, the 1943 bicycle output was expected to have to be around 2,500,000 units to meet demand.[20]

The following month, the WPB issued an order intended to stop the "terrific rate at which bicycles have been going to people who don't need them." It was printed on the front page of *The New York Times*, and it was brutal. The order immediately froze the "sale, shipment, delivery or transfer of all new adult bicycles"—even if already ordered, paid for, and in transit to the retail consumer. Future sales priority would be given to defense workers and essential civilian needs. It also redefined adult bicycles as being more than 19 inches from the center of the crank to the top of the saddle pillar post. A week later the Board defined an adult bicycle to mean any that was more than 17 inches from crank center to the top of the pillar post. The average American adult had apparently shrunk three inches in four weeks.

In some respects, however, the WPB had slammed the door after the iron steed had bolted. With the severe restrictions on private motoring, New York cycle dealers reported that virtually all new adult bicycles, and most used ones, had been sold prior to the freeze order anyway. Whereas sales in winter months were normally abysmal, the city's dealers had not been able to keep up with demand in early 1942. In fact, distributors redirected machines from areas of light demand to shops near commuter centers, where they were heavily used for getting to and from railway stations.

In May 1942, the OPA announced that it would ration bicycles, and in June it set the price of the plain, basic "war model" (the only one produced) at $32.50 on the east coast, slightly higher in the west. To get one, a worker had to be contributing to the war effort or public welfare, and show that he had to travel quickly or often in delivering goods or messages, at least three days a week, and could do so faster on a bicycle than by walking or using public transportation. Also, anyone having to walk more than three miles a day to get to work, or spend more than one and a half hours in travel, was eligible, providing that the bicycle saved at least 30 minutes.

It took a while for the rationing system to satisfy demand. In the interim, the public got their cycles wherever they could, and by whatever means. In Columbus, Ohio, at a police auction, "children watched pop-eyed" as adults bid up to $33 for 54 used bicycles that normally would have gone for only a few dollars. More nefariously, within days some dealers were taking apart new bicycles and using the parts in second-hand machines, where the rationing restrictions did not apply. Others simply shortened the seat pillar to just under 17 inches by cutting the top off, so that it became a "child's" bike, which was not subject to rationing, and then using extended seat inserts. The OPA soon stopped it by rationing all new bicycles and parts, whether disassembled or incorporated in used machines. And in response to complaints, it ordered dealers to sell cycle tubes separately. Some had required customers to purchase the tire and tube as a unit.

In mid-August, 1942, the War Production Board stopped all bicycle manufacturing. It felt that there were enough machines and parts to go around, if properly allocated. The WPB simultaneously tightened up the list of eligible purchasers. Even those still qualifying for bicycles were denied them if they had usable tires on their cars and sufficient rationed gasoline. In a nutshell, "The OPA does not intend that car owners should purchase bicycles to save their tires or gasoline for nonessential driving." Bicycle manufacturers' facilities were converted to making machine gun tripods, searchlights, incendiary shells, and aircraft parts.

There was one last matter that had to be resolved. When used bicycles were subjected to rationing restrictions, no ceiling was placed on the price dealers could charge. Being the only ones available with "brightwork"—all that fancy chrome, silver, and other accessories not on the war models—prices soared to what many thought were exploitative levels. Consequently, in December, 1942, the OPA set the maximum price of any rebuilt bike at

Displaced civilians in northern Europe, October 1944

eighty percent of what it had cost new in February, with an absolute upper limit of $32.50, the same price as the war model.

In October 1944, the OPA formally ended cycle rationing because so few new machines were left that it was not worth the administrative effort. On May 22, 1945, with the end of the war in sight, the War Production Board lifted all restrictions on the manufacture of bicycles (along with lawn mowers, fly swatters, and carpet cleaners). The agency warned, however, that shortages of materials would retard output for awhile. Most Americans did not care. Within 24 hours of Japan's surrender, in August, gasoline rationing ended and they climbed back into their cars. The number of motor vehicles shot from 26 million in 1945 to 40 million in 1950. Two years later, the new president, Dwight Eisenhower, began to lobby for the development of the greatest highway system on earth. A new bicycle was now among the least of adult Americans' interests.[21]

8. The Bicycle in Vietnam

Harrison Salisbury: "I literally believe that without bikes they'd have to get out of the war."
Senator William Fulbright: "Does the Pentagon know about this?"
U. S. SENATE FOREIGN RELATIONS COMMITTEE, OCTOBER 1967

Vietnam was wracked by fighting from the time of the Japanese occupation in 1940, until 1975. During that period the safety bicycle saw its greatest ever, most spectacular, and longest running, role as a military logistics tool—three-quarters of a century after its development. When it was all over, in 1975, the Vietnamese had been fighting the Japanese, French, Americans, or one another almost continuously for 35 years, longer than all the major wars of the century combined. The country was left economically depleted, physically scarred, and millions dead or permanently maimed. In the process, Japan lost its dream of a Greater East Asia Co-Prosperity Sphere, the French lost a major element of their empire, and America lost its first war. Throughout it all rolled the bicycle.

For many Americans, the Vietnam war was confusing. Among other things, it was about the capabilities and limits of technology. Appropriately, as the first television war, it used the most sophisticated reporting technology available. But some believe that the cameras, the most common source of war "information" for many, ultimately revealed little of the underlying problems and nature of the war. They focused, one sidedly, upon the spectacular: napalm attacks, bombing patterns seen from on high, soldiers firing at unseen targets, and the close up results of burnt villages, fleeing civilians, dead livestock, and body bags being loaded onto helicopters.

In contrast, there was little nightly footage from the other side of Vietnamese porters carrying 50-pound loads down steep slopes, or climbing out of holes immediately after heavy bombing attacks to continue pushing heavily laden bicycles along muddy

Vietnamese porteers carrying laden bicycles up a hillside

roads. Even if the networks had the footage, such scenes would have been too mundane for prime time news, night after night. Yet, those simple, unspectacular actions, repeated innumerable times around the clock, coupled with the political determination to drive foreigners out of their homeland, was where the war was ultimately won.

During it all, some portrayed the bicycle as a symbol of the conflict: peasant level technology against sophisticated weaponry and "war management" techniques. Many dismissed the comparison, and others wrote derisively of "that wish fulfillment of parlour-pink intellectuals, the soldier peasant wheeling his bicycle to victory." Most gave it no thought at all. But whether the bicycle was more than a symbol, or less than the key to victory, it certainly provided one basic lesson for all who wished to observe, think, and perhaps learn: the device was still a component of the military scene. It had not been made obsolete, but was only overshadowed by other technological developments. Because those other technologies cast such *dark* shadows, however, the bicycle was left nearly invisible. It is no wonder that, in the end, when the television-informed public was trying to decide what it was all about, "only the commercials made sense."[1]

The First Indochina War

Vietnam is a relatively narrow 1,100 mile arc paralleling the coast of the South China sea. Along the northern and western borders with Laos are the rugged Annamite mountains, which form Vietnam's central highlands. As the Australian journalist Hugh Lunn discovered, it is difficult for people without experience in the area to imagine the extent of isolation of, and difficulty of access to, the small mountain villages. In striking contrast is the extensive, flat, densely populated Mekong River delta in the south, and the Red River delta, about Haiphong, in the north. Throughout Vietnam, with its hot, wet, tropical climate, is "dense spontaneous growth," as Bernard Fall understatedly described it. Most simply call it jungle.[2]

After the occupying Japanese were defeated in 1945, the French again took control of Vietnam. The communist Viet Minh were determined to drive the French from their homeland, and during the late 1940s developed their tactics and strategies. Politically, Ho Chi Minh persuaded the essentially peasant society that it was a struggle to drive out the foreigners and return the land and social system to themselves. It did not take much convincing in many quarters. The key figure militarily was Vo Nguyen Giap. He developed the technique of many small scale actions, no single one being important, but cumulatively raising the enemy's anxiety level and destroying his self confidence. Giap proved a genius at logistics, able to "move men and supplies around a battlefield far faster than anyone had any right to expect," and learned how to work with villagers without being betrayed by them. In a prolonged struggle during which it was not always certain who was on who's side, that was no small talent.[3]

The French forces were well equipped. They had good armored vehicles, total control of the air, and patrolled the watercourses with the *dinassaut*, heavily armored flotilla that were highly effective. However, the Viet Minh created a war without fronts by fighting in small units throughout the countryside, and avoiding major battles and urban areas. Ultimately, the French never could overcome the Viet Minh's penetration and control of the jungle. In 1950 Giap finally undertook his first major offensive against the French. Although he suffered severely, the French losses that October—6,000 troops killed or captured—constituted the greatest colonial defeat since Montcalm died at Quebec.[4]

In 1953, a magazine published by the French Information Services in Indochina commented that "the comparison between the supply of a well-equipped [French] squad mounted in a half-track and that of a [Viet Minh] squad trotting through the jungle, each man with a four-day ration of rice in a bag slung about his neck, would be ridiculous had it not proved so tragic [for the French]." After seven years of war France had lost 74,000 troops, with another 190,000 bogged down. General Henri Navarre went to Vietnam to assess the situation in May 1953. "It struck me immediately that there was no possibility of winning the war in Indo-China. The Viet Minh had gained a considerable advantage over us in mobile forces." He proposed "that we find an honorable exit from the war."[5]

General Navarre decided to take on the Viet Minh at Dien Bien Phu, in the extreme west of the country. It was located in a valley on the main route to Laos, where a crucial Viet Minh supply route from China linked up. Navarre wanted to draw Giap into a set piece battle at this crucial transport junction, believing that the Viet Minh did not have sufficient transport to bring in the needed food, weapons and other supplies to win a major confrontation in the isolated setting. If Navarre could defeat them, it would allow for strong negotiations by the French during their "honorable political" withdrawal from the country.

On November 20, 1953, the initial French forces arrived in Dien Bien Phu. Within a week 10,000 soldiers were in place and another 5,000 in reserve. In early December Ho Chi Minh and Giap decided to take on the French. Giap proposed putting 50,000 regular troops around Dien Bien Phu, supported by tens of thousands of porters and workers who would cut out new jungle routes along which the soldiers could march, and who would carry in weapons and supplies. The French Air Force's effectiveness would be limited by having to operate at maximum range against Vietnamese troops hidden under a thick canopy of trees.

The Bicycle and Dien Bien Phu

Dien Bien Phu became, above all, a battle of logistics. In the final analysis, the French were at a great disadvantage, having to fight far from their main bases, in a valley surrounded on all sides by higher ground held entirely by the enemy, and totally dependent upon air support. Critically, the French and their United States advisers did not consider that the airfields of Dien Bien Phu could be severely restricted, let alone closed down by the Viet Minh. Thus they did not stockpile sufficient supplies within the fortress before the battle began, and established only a small field hospital on the assumption that all wounded would be flown out.[6]

The gravest error made by the French was in underestimating the ability of the Viet Minh to bring up heavy artillery and supplies for their troops. The French expected to have only mortars arrayed against them, but in fact faced 144 pieces of heavy artillery, some thirty 75-mm recoilless cannon, 36 heavy flak pieces, and a dozen six-tube katuysha brought up in the last days of battle. Much of it was dragged to Dien Bien Phu by relay teams of men roped to the artillery, pulling it literally inches at a time in some places.[7]

The 49,500 Viet Minh soldiers who eventually served in the surrounding hills were supported by an organized supply effort that was a truly remarkable achievement, administratively and physically. In the northern delta regions, several thousand villages sent huge quantities of rice and tens of thousands of porters to support the effort. Additional rice, artillery, and heavy ammunition were both trucked and portered hundreds of kilometers south from the Chinese border. The supply lines converged on Son La, the nodal supply point 165 kilometers east of Dien Bien Phu. The route from Son La to Dien Bien Phu was divided into three sections. The first was a 100-kilometer stretch to Tuan Giao; the second was a 47-kilometer segment along an old colonial road that had to be rebuilt; and the third was a short but difficult 18-kilometer run, under enemy observation, that led directly to the rear store houses of the front line divisions. Along each section, transshipment points and river crossings were established. These were manned by supply personnel who received, sorted, packed and earmarked stores for specific units at the front. Work brigades and engineers repaired the roads and bridges damaged by French aircraft attacks.[8]

The key to the logistics system was the combination of transport modes, built around the backbone of the greatest military cyclist transport effort in history. Although the Viet Minh used some 600 Russian Molotova 2.5-ton trucks, augmented by hundreds of cars, and thousands of rafts, sampans, and mountain ponies, the mainstay of the network was tens of thousands of "tough bicycle-pushing human supply columns." When surfaces were too steep, muddy, or otherwise unsuited to cycle porters, goods were transferred to the backs of humans or animals. The number of porters and workers used, and amounts carried, will never be known for certain. Western estimates suggest that at least 200,000 porters were employed. In various Vietnam sources the number of porters has usually been put at around 260,000. Added to that would be the work brigades keeping the routes open. The tens of thousands of men, women and children were organized in thousands of villages by political commissars.[9]

Bicycles, with their great carrying capacity, were particularly effective on the narrow roads and tracks in the dry season. The photographs of porters pushing heavily loaded and camouflaged machines along trails, over bridges, and up and down slopes of varying grades and surface conditions, suggest the tremendous effort that was required individually, and the cumulative effect of the columns in getting massive amounts of supplies to their destination.

A brigade leader from Pho Moi village, bicycle repairman Ding Van Ty, said that "We had one day to make preparations. First our bicycles had to be turned into *xe tho* [pack bikes], with the crossbar capable of carrying 200 kilos or more. We had to strengthen all the parts . . . We camouflaged everything with leaves and moved at night." Typically, the seat was removed and a rack of some sort placed over the back wheel, either a standard metal one, or a wooden or bamboo structure lashed in place. This provided an extended line along which could be hung bags, boxes, or other items. Goods were tied on by rope, strips of inner tube or whatever else was at hand. The frame was often strengthened by adding metal, wood or bamboo reinforcing struts. Reinforced front forks were common on the third world bicycles.[10]

Once loaded, it was usually impractical to walk close enough alongside the bicycle to use the normal handlebars for

Cargo bicycle with reinforced fork, steering extensions and vertical handhold

Vietnamese porters supplying Dien Bien Phu

steering. Hence a stick was commonly lashed to the handlebars, which extended out far enough to allow the porter to hold it and steer comfortably. As well, another stick was lashed to the frame or load somewhere near the seat area, or stuck into the vertical seat tube. The porter used it both to push the bicycle along, and to hold it back on downhill slopes.

The weights that the porters pushed were immense, estimates ranging up to 600 pounds. However, the most common weight cited, both by Vietnamese and westerners, is about 200 kilograms, or 440 pounds. Numerous photographs showing bicycles laden with large bags of rice, huge wooden crates, and other items, would tend to support that figure. The great weights carried is evidenced by a Vietnamese porter's bicycle now held at the Australian War Memorial. At the point where the frame tubes are welded into the lugs, near the handlebars, the tubes are crinkled like paper, indicating the extremely large forces exerted on the joints. It is testimony to the structural strength of the machine.[11]

Tricycles were occasionally employed, but not extensively. They are widely used in urban areas as taxis and cargo carriers, can hold great weights, and are particularly useful in carrying large, bulky objects. Their principal disadvantage is the need to negotiate three wheel tracks simultaneously along a surface without one or another wheel hitting an obstacle or becoming bogged in a hole or rut. On rough, narrow tracks, especially in wet weather, they are essentially impractical.

Bernard Fall attempted to calculate the tonnage moved by the Vietnamese logistics system. He provided no detailed figures for supplies of local Vietnamese origin, but concluded that from China alone, some 8,300 tons were shipped to Dien Bien Phu, including petroleum products, ammunition, spare weapons, and rice. The immense amount of materiel that was moved by hu-

Several heavily camoflaged cargo cyclists

man energy was simply not anticipated by the French. The overall supply system was aptly characterized as a "human serpent" that came up from the plains and wrapped its coils around the French garrison.[12]

The Viet Minh prepared for three and half months before attacking. When they did, on March 12, 1954, they opened up with an artillery barrage from much closer than expected, far heavier and more devastating than anticipated, and followed by intense ground assaults. The Viet Minh quickly took outlying French positions, and two weeks later closed the central airstrip. From then the battle became a siege. About 120 tons of supplies were parachuted on to the French encampment daily, but about 20 tons were destroyed or fell into Viet Minh hands.[13]

The Viet Minh continued to receive supplies, despite French efforts to stop them. Although the Vietnamese supply route was relatively well known, and ended at a very specific point, the French Air Force had only a small amount of aircraft and bomb tonnage available. As well, the Viet Minh camouflaged part of the route, in places tying the tree tops together to create tunnels over the roads. In the end, the French air force failed to hamper supply operations to any significant degree. Whether more could have been done has been argued. However, it must be remembered that, barely a year earlier, the United States Far Eastern Air Force had failed to bring the Chinese and North Korean supply systems to their knees, despite Operation Strangle's year-long round-the-clock interdiction effort by hundreds of American heavy and medium bombers.[14]

French aircraft could not overcome the Viet Minh's four-to-one artillery superiority. On May 7, 1954 the French surrendered. Some 8,000 Viet Minh and 3,000 French soldiers had died. But for many of the French, the suffering was only beginning. Over the next 60 days some 8,000 survivors were marched several hundred miles to prison camps in the Red River delta, and half died on the way or in camp. For the soldiers on both sides, Bernard Fall's book title, *Hell in a Very Small Place*, aptly describes Dien Bien Phu. In terms of length and the number of men involved, it was not a major siege by world standards. Rather, its significance was historical, in marking the end of the French presence in Vietnam. As well, it offered the most impressive demonstration in history of the logistical capabilities of the bicycle.[15]

The Second Indochina War

Post-Dien Bien Phu politics are beyond the scope of this book, and are the subject of many arguments and interpretation. In summary, the Geneva Accords of 1954 left Vietnam divided, but called for joint elections within two years to decide reunification. However, the United States backed the South Vietnamese leader, Ngo Dinh Diem, who refused to support a national election. Instead, he held a plebiscite in the south and was elected "chief of state." Unfortunately, his social and land redistribution programs did not go well and he lost much support over the next few years.[16] In the meantime, the communists began efforts to unify the country. In particular, they concentrated upon identifying the Diem government with the United States, thus converting the political situation into yet another nationalistic struggle, in which the Americans simply replaced the French. It was a message readily accepted by many southern Vietnamese peasants, who suffered under Diem. In 1958 Diem coined the pejorative term, Vietcong, to describe guerrillas then operating in the south. In 1960 the political arm of the Vietcong, the National Liberation Front (NLF), was established in many southern villages.

The Americans responded by increasing military and other aid to South Vietnam. In 1961 the first United States helicopter units were sent. By the end of 1962 there were 12,000 "advisers." Eighteen months later 125,000 American combat troops were committed. The buildup quickly escalated, and at its peak the war occupied 543,000 American military personnel on site, along with another 100,000 allies. Several million Vietnamese were directly involved in the fighting on both sides, and civilians in both North and South Vietnam were severely affected.

Thus, in simple terms, evolved the "Vietnam War." Among those who remembered that the French had already fought the Vietnamese for nine years, it was considered more accurate to refer to it as the Second Indochina War. First, Second ... for the Vietnamese they were but phases of what ended up being the longest war in modern history. And, short of nuclear weapons, the second phase would witness the application of the most sophisticated military technology yet devised, see the helicopter gunship come of age, and experience the greatest concentration of explosive power ever delivered. But in the end the victors were still using bicycles, as they were in the beginning.

The United States was facing essentially the same enemy, with the same leadership, over the same terrain, as had the French, and it would have seemed logical for the Americans to turn to them for advice. The French were certainly willing to offer it. The French military attaché to Saigon in the mid-1960s was "handpicked by the French government because of his exceptional knowledge of the English language, and his distinguished record in Indochina and Algeria. He was told to help the Americans in whatever way he could." However, during the first 18 months of his assignment, the only American who visited him to ask about the war was an American defense contractor of French origin. In fact, many Americans felt that they had solid grounds for ignoring the French. Many U. S. advisers had disagreed with the French about how to deal with the North Vietnamese in the First Indochina War, and the French had lost. If only they had listened .....[17]

David Halberstam wrote in *The New York Times Magazine*, in 1962, that the American military in Vietnam tended to be too critical and "almost patronizing about the earlier French effort here, and it would probably behoove them to remember . . . that the French were not exactly naive in the war . . . and that it has yet to be proved that the present effort is really combating the enemy where he lives any more than the French effort did." Bernard Fall observed that the basic lesson to come out of the French involvement was painted in large letters in the halls of the French guerrilla warfare school in Vietnam, and appeared every month on the first page of its monthly magazine: "Remember—the enemy is not fighting this war as per French Army regulations." The word *French* would now have to be replaced by *American*.[18]

In essence, there was a widespread belief that once the Americans engaged the Vietnamese communists, it would be a very different matter, technologically, from when the French fought them, that American might would do what the French

Technology, Perceptions and Experience

could not. For example, the award-winning Australian journalist, Hugh Lunn, who covered the war for Reuters in the late 1960s, admitted that in 1965 he had argued with Chinese students in Peking that the Americans "could win anytime they wanted; all they had to do was to get serious and start using jet bombers and other big equipment that had not yet been brought into play." At Dien Bien Phu, for example, there was an average of only 155 combat sorties and 281 supply missions a week flown in support of the French troops. In contrast, by the beginning of 1966 the United States was sometimes flying in excess of 25,000 sorties a week.[19]

General Westmoreland considered that "the most spectacular development was the coming of age of the helicopter. It saved innumerable lives through air evacuation. It gave us a battlefield mobility that we never dreamed of years previously." It was possible to move, within hours, huge amounts of men, materiel, and firepower across terrain and over distances that would take the enemy weeks. Indeed, for many the helicopter came to be the symbol, sound and memory of the war. It augured for a total revision of the concept of mobility in the jungles of southeast Asia. From that optimistic perspective it really did look as if it would be a different war from that which the French had fought.[20]

But the American attitude and approach to technology was often based upon an exaggerated idea of what it could accomplish. Michael Herr, in his award winning book *Dispatches*, referred to the ability of the gatling gun-armed helicopters to fire 300 rounds per second. Such power, awesome enough in its own right, was described by him as capable of putting "a round in every square inch of a football field in less than a minute." That is wrong. There are 6,480,000 square inches in a hundred yard by fifty yard football field. Three hundred rounds per second for sixty seconds is only 18,000 rounds. Thus, only one quarter of one percent of a football field can be covered in a minute. It would take six hours to fill every square inch. Herr miscalculated by a factor of 380. Literary hyperbole? Perhaps. But it more likely symptomatic of the problems even thoughtful Americans can have in coming to grips with the capabilities and limitations of technology. And no form of technology is invincible—by 1968 the United States had lost several thousand helicopters.[21]

Giap and Ho Chi Minh approached the war from a different perspective. They focused first upon ensuring that the Americans did not succeed in making it appear to be a fight between northern and southern Vietnamese. They made sure that it was viewed locally as a war between the Vietnamese people and America. It worked. As one Vietcong leader, Truong Nhu Tang, observed, an American writer once disdainfully declared to him that "the average guerrilla couldn't have told dialectical materialism from a rice bowl." Tang conceded that, by and large, that was true. But, he countered, "as far as most of the Vietcong were concerned, they were fighting to rid the country of foreign domination—simple motives that were uncolored by ideological considerations."[22]

Militarily, Giap again utilized the basic principles used against the French. He undertook a strategy of protracted war to create opportunities and gain time, and used North Vietnam as a sanctuary and source of reserves. Many Americans expected an eventual repeat of the Korean scenario, that is, an invasion by the People's Army of North Vietnam across the DMZ. But it never came. Instead, Giap spread his men across the southern Vietnamese countryside, extending "the elusive front of guerrilla warfare in every direction." And Giap had a very large battlefield across which to spread. North and South Vietnam, along with eastern Laos and Cambodia (where both sides fought, illegally and unofficially, for some years), were greater in cumulative area than the state of California. Analysts of the war believe that Giap's ability to keep United States and ARVN forces dispersed across such a large area was the critical key to maintaining the strategic initiative.[23]

In terms of technological sophistication and firepower, Giap could not match the Americans. But he attempted to neutralize their advantages by moving and fighting at night and in the rainy season, using ambushes at close quarters, and avoiding routines that could be anticipated. On the other hand, it was by no means a simple case of a peasant force against a technological giant. The People's Army of Vietnam (PAVN), the fifth largest in the world, was one of Asia's best; it was reasonably well equipped and became more so as the war dragged on; and it had a ubiquitous network of agents which was used effectively. The latter meant that many American analysts "could never agree on what constituted military enemy and what constituted non military enemy. By the enemy's own definition, there was no such thing as "non military"." Thus, a central characteristic of the Vietnam War was that no one on either side—and especially the Americans—could ever be absolutely certain who was on which side."[24]

Giap had not just altered the rules of the game, as Americans understood them, but had substantially changed its nature. The traditional American military orientation—combat with an identifiable enemy—was thrown into disarray by the confused mixture of ideological, political and military elements in which soldiers and civilians dressed alike. The result for many allied soldiers (and politicians) was bewilderment at why they could not come to grips with an enemy which, as one Australian colonel described them, was using "bicycles as utilities [pick up trucks] and elephants as five-ton trucks."[25]

The Ubiquitous Bicycle

In the Second Indochina War the bicycle again played a significant transport and logistics role. While there was no single, spectacular focus, such as at Dien Bien Phu, anyone who served there could not help but be aware of the millions of cycles in daily use which could, and did, serve equally well for civilian and military use. A western perspective on that use was offered by Wilfred Burchett, an Australian journalist who covered the First and Second Indochina Wars, including Dien Bien Phu, from the communist side, and was widely published in European newspapers. The first westerner to travel the Ho Chi Minh Trail, Burchett was trusted implicitly by the communist Vietnamese regime, to the extent that the Australian government deemed him to be aiding the enemy and revoked his passport. In his book, *Vietnam: Inside Story of the Guerrilla War*, Burchett reported on his clandestine travels about the southern Vietnamese countryside in late 1963 and early 1964 with guerrilla cadres. He combined the eye of a western journalist with the guerrillas' perspective. The book simultaneously reflects sympathy with their cause, and a foreigner's astonishment at their achievements under difficult circumstances.[26]

Burchett, wearing parachute camouflage, with his typewriter on the back rack

Burchett highlighted the flexibility of guerrilla travel by covering over 500 miles by bicycle in just a few months, in addition to extensive walking and riding in sampans. His first cycle, French-made, had the frame and both wheels stamped with clasped United States-Vietnamese hands under the Stars and Stripes, with a notice that it was a "gift of the people of the United States." On his first day he bicycled 25 miles before arriving at a quiet, brown stream where he rested. That evening two sampans arrived, one with an outboard motor towing the other. They abandoned their bikes to climb aboard the boats. His guides changed from region to region, and at various rendezvous points yet more bicycles awaited them.

The riding conditions could be difficult: "A narrow, winding trail, never more than three or four yards straight, with roots and snags everywhere; tiny stumps where the undergrowth had been slashed close to—but not level with—the earth, jabbing at your pedals and ankles; overhead creepers waiting to strangle you while you are looking down to avoid a stump; trellises of bamboo banging at your head no matter how low you bent over the handle bars; a multitude of spikes reaching out to rip your shirt and flesh."[27]

After riding in such conditions for hours, Burchett was grateful for anything which resembled a highway. On exposed roads, a valuable characteristic of the bicycle became immediately apparent: "It was better than being in a jeep because with the silence of bike travel we always had plenty of warning of approaching planes and could pull into the undergrowth." Once, when pushing through heavy soil in a clearing, they had to shoulder the bicycles and run, as American planes circled looking for a target.

He found that when there were several bicyclists in a party, they inevitably got strung out along the narrow paths. When they came upon a hub of tracks radiating out in several directions, the local guide would indicate to those behind which track was to be followed. It was critical, because they often passed within one or two miles of an enemy post, and only a few minutes of cycling down a wrong path would put them in great danger. He was occasionally warned to follow the next bicycle "very precisely" because the road had been mined by the guerrillas with spiked or explosive traps.

One thing fascinated Burchett.

> "I never failed to be surprised when, for instance, cycling along a tiny crack through dense jungle, the cook would suddenly lean over and, without slackening speed, extract a chicken for my lunch, its legs tied together with a bit of jungle creeper. I never understood how such rendezvous with chickens were arranged and even less the mystery of how the cook knew behind which bush it had been deposited."

Burchett's most difficult times were when he had to cross the innumerable swaying bridges "suspended from cables of jungle creepers, with woven matting laid across lengths of bamboo as a footway and creeper handrails on each side." At least they provided a flat and reasonably wide surface. The worst crossings were those spanning a gap with "a single perfectly round tree trunk of giant bamboo, tapering from one foot down to about six inches and with a single, usually loose creeper at about waist level. This was useless for support but did help in correcting one's balance." The whole matter was complicated by having to carry his bicycle across on his shoulders.

Most of the Vietnamese wore "Ho Chi Minh" sandals,

with the soles made from tires and the thongs from inner tubes. Burchett noted that the only change in them from the First Indochina War was that "Goodyear has replaced the French Michelin as the main supplier." He found the sandals dangerous when crossing the single-log bridges, which were often humid or greasy from the recent passage of muddy feet. "One always tackled these barefoot, walking in a slightly splay-foot fashion, to give each foot its maximum grip over the width and curve of the trunk. I only discovered this trick by carefully watching the tribespeople. At first glance they seemed to skip effortlessly across but in fact they placed their feet very carefully." He learned to feel with his feet and watch the far bank, because looking down was unnerving when there was a long drop.

After traveling about the rural areas for some time, it seemed obvious to Burchett that the communists could cut many Saigon-held roads and take control of many provincial and district centers, but had not done so. It eventually became clear that the roads were the guerrillas' supply lines, and the towns their distribution centers. To capture them would simultaneously draw South Vietnamese Army attacks and paralyze their own supply system. In early 1964 the NLF was not yet ready for such action on a large scale.

A Vietcong view of the daily use of the bicycle during the war was provided by Truong Nhu Tang, in his book, *Journal of a Vietcong*, written after spending six years living and traveling in Vietnam, Laos and Cambodia. Tang and his comrades traveled light, for they often had to move fast. Soviet intelligence trawlers monitoring American flights advised the communists when American air raids were coming. "Often the warnings would give us time to grab some rice and escape by foot or bike down one of the emergency routes. Hours later, after the carpet bombings, "we would return to find, as happened on several occasions, that there was nothing left. It was if an enormous scythe had swept through the jungle, felling the giant teak and go trees like grass in its way, shredding them into billions of scattered splinters ... It was not just that things were destroyed; in some awesome way they had ceased to exist."[28]

The guerrillas carried weapons and ammunition, but

under constant threat of attack, kept their personal items to a minimum. "Two pairs of black pajamas, a couple of pairs of underpants, a mosquito net, and a few square yards of light nylon (handy as a raincoat or roof) were all that a guerrilla owned." They carried the basic food staple, rice, in long tubes of rolled cotton (known as "elephant's intestines") slung across their back. The rice ration was the same for everyone, 20 kilos a month, and was eaten at about nine in the morning and four in the afternoon. Although sometimes supplemented by a bit of dried fish or meat, it did not go far, and left them often in a state of semi starvation.

Once, under tremendous pressure, Tang and his colleagues were forced to move with little letup for five days to avoid being cut off by South Vietnamese forces,: "As we walked, the rains, typical for that time of year, poured down continuously, turning the red Cambodian earth to a sticky clay that sucked at our rubber sandals, until the last of them had been lost or discarded. Barefoot, pants rolled up above our knees, we shuffled ahead in the ankle-deep mud, each step an energy-draining struggle. Those who had bicycles abandoned them beside the muck of the trail."

Until 1970 the usual means of transport continued to be the bicycle and walking. However, Tang recalled that "in the winter of 1969-1970 the whole country was inundated by an invasion of Japanese motorbikes. In one way or another, these bikes made their way out from the cities and into the hands of even the most remote country people, who would then smuggle them to the guerrillas." From then on, "we had a regular supply, not just of Hondas, but of typewriters, radios, cigarettes, and a variety of other goods, bought or stolen by the peasants from the local Saigon army forces."

The bicycle's widespread use in the growing war led to it becoming an occasional topic of discussion. In October 1967, Harrison Salisbury raised the matter in United States Senate Foreign Relations Committee hearings chaired by Senator William Fulbright. Salisbury testified that communist supplies were bicycled to South Vietnam in such large amounts that "without bikes they'd have to get out of the war." It was a reiteration of a message he had written in his book and on the front page of *The New York Times* ten months earlier. "If by some magic weapon all the bikes in North Vietnam could be immobilized, the war would be over in a twinkling." The machine was "as essential to the North Vietnamese as the auto is in Los Angeles."[29]

Senator Fulbright responded with two questions: "Why don't we concentrate on bicycles instead of bridges? Does the Pentagon know about this?" Those in attendance laughed, and the remark was reported in a lighthearted manner both in the United States and abroad. In London they described "bombing bicycles instead of bridges" as the new United States plan to force Hanoi to the peace table.[30]

Unknown to Fulbright and Salisbury, the Pentagon did know about it. Taking a lesson from the French experience, they had commissioned a study on military bicycle troops two and a half years before, on March 1, 1965. It was carried out by the Battelle Memorial Institute's Remote Area Conflict Information Center, in Columbus, Ohio. The report, *Bicycle Troops*, was written by the late R. S. Kohn. In an interview eighteen years later, he said that it was but one of many studies conducted by the Institute at the time. He had not been high enough administratively to be in on the initial contract talks, and knew nothing about the circumstances leading to its commissioning, but had the impression that the study was part of an overall effort to better understand mobility. The officer requesting the report, Colonel B. F. Hardaway, was chief of the Advanced Research Projects Agency Research and Development Field Unit—Vietnam. The introduction to the report noted that "interest in the employment of bicycle troops is emerging once again, this time in Southeast Asia where the road network is inadequate for motorized transportation, but where paths and dikes may provide an acceptable avenue for bicycle movement." That comment was taken verbatim from an earlier joint United States-Thai report prepared in Bangkok.[31]

In the covering letter to Colonel Hardaway, it was pointedly noted that information on bicycle troops was highly fragmented and widely dispersed, and that little had been written since World War II. Most of Battelle's information came from the first 25 years of the century. Their researchers relied essentially upon American sources, plus a few foreign holdings in the Library of Congress. Kohn felt that Japanese archival records on the use of the bicycle in Malaya might prove especially valuable. However, Battelle inquiries to United States Military Attaches both in Japan and other countries known to have used bicycle troops elicited not a single response.

In summary, if Colonel Hardaway's intent was to get information to help the Pentagon assess the use of bicycles in Vietnam, and develop countermeasures, he was out of luck. Despite the introduction's reference to the importance of the Japanese use of the machine in Malaya, the body of the report contained only a single short paragraph on Malaya from one reference. With respect to previous wartime cycle use in Vietnam, there were two references to Jules Roy's book on Dien Bien Phu, totaling seven lines (and one bit of information was used in both references); a single sentence noting that "In Vietnam today terrorists on bicycles are active, and bicycle handlebars have been used by guerrilla sympathizers to smuggle contraband past roadblocks"; one example of a Vietcong bicycle booby trap, with an illustration; and six and a half lines about an anonymous "British civilian" (actually the Australian, Wilfred Burchett) who used the bicycle in Vietnam. That was it.[32]

What the report essentially provided was a short historical survey of early military cycling, plus 71 pages of tables, translations, and illustrations mostly from or about German cycling experiences in World Wars I and II, drawn heavily from *Die Radfahrtruppe*. Most of the remaining quotes were from the 1890s, and particularly those of Lieutenant Moss. It is not surprising that the report's security rating was "Unclassified," and few ever heard about it during the Vietnam era.

Kohn said that a major problem he faced in preparing the report was that he was denied access to many records, such as World War II OSS files on how to put explosives in bike frames. As Harrison Salisbury laughingly commented, when hearing of the restrictions placed on Kohn, it would be like the Pentagon to commission a study, then deny the writer access to pertinent information. Salisbury thought the bicycle study was in all likelihood a typical product of the Robert McNamara era, in which the Secretary of Defense would say something like "let's cover every technical possibility," and numerous studies would follow. It is not known what the Pentagon's overall conclusions were as far as the bicycle was concerned.[33]

The Ho Chi Minh Trail

The Ho Chi Minh Trail became the most critical and famous transport link of the war. It was a network of paths, roads, streams, and rivers running down the spine of the Annamite mountain chain of eastern Laos, through some of the most inhospitable terrain and impenetrable jungle in the world. Paralleling and including the border areas of Vietnam, Laos and Cambodia, it provided a route for sending Viet Minh infiltrators and, later, regular troops from North to South Vietnam, and for keeping them supplied. Its construction and maintenance, and the very act of traveling it, were impressive examples of dogged human endurance, perseverance and organization. Bicycles were used both along the trail, and to disgorge men and materiel from the trail into the South Vietnamese countryside.

Early trails were established by mountain tribesmen who had long inhabited the area, and by traders who used the network for millennia for coolie caravans moving gold and opium between China and Southeast Asia. The Viet Minh began using the isolated routes as communications links during their war against the French. In 1959 Viet Minh residing in the north began traveling southward down the trail in increasing numbers, supported by porters carrying modest amounts of arms, ammunition and other materials.

The trail was not a single track, but a complex web of jungle paths, and was a difficult ordeal for travelers. They required several months for the trip, each person carrying the supplies and equipment he needed. "In the suffocating humidity they climbed and hacked their way through the thick foliage of the mountain jungles ... in one of the world's most hostile environments." They had to cope with malaria, dysentery, fungus infections, clouds of mosquitoes, and the highly venomous krait snake. "Ten kilometers or so a day [six miles plus] was all they could expect to advance, and those who became too sick to walk were simply left in primitive shelters, which had been constructed by even earlier advance parties."[34]

In 1963, as United States military aid increased, the com-

Vietnamese crossing a rudimentary bridge

munists realized that in order to support the NLF in South Vietnam they would have to send regular Vietnamese forces south along the trail. General Chu Huy spent four months investigating the feasibility of expanding the trail system. In October 1964 (sparking the intervention of United States combat forces), the first PAVN tactical unit from the north headed south along the trail. It was still an arduous journey along a primitive route that could not handle large numbers of troops. Bicycles and ponies were used on portions of the trail to carry supplies, but were not practical over much of it. Across the 7,000 foot passes and along the mountainous heights it was still essentially a narrow footpath, with steps cut into steep slopes. Over many streams, Burchett noted, flimsy, meter-wide bridges with bamboo foot planks and ropes for handrails were strung across temporarily and then dismantled, to avoid detection.[35]

In 1964 the decision was made to enlarge the trail into a truck route, with antiaircraft batteries to defend it and engineers to maintain it. The following year the American navy began Operation Market Time, which dramatically cut the southern movement of communist supplies along the coast. In 1965 an estimated 70 percent of North Vietnamese supplies moved south

A Vietnamese two-bicycle ambulance, rigged to carry four sitting wounded and two stretcher cases

by sea; two years later it was down to only about ten percent. The difference had to be made up along the Ho Chi Minh Trail, and construction efforts were intensified. By 1966 the trail was a relatively well marked series of roads, handling heavy truck traffic, with underground support facilities, including supply caches and hospital and medical services.[36]

The Ho Chi Minh Trail was the backbone of a system with many branches. Supplies and troops exited at three major points: the A Shau Valley, the Ia Drang Valley, and War Zone C. It was from such points (and many other smaller branches) that the bicycle and other transport modes were particularly relied upon to deliver supplies to troops dispersing within southern Vietnam. By 1967 the trail system had become the key to the war's progress. Whereas between 1959 and 1964 North Vietnam had infiltrated only about 30,000 personnel into South Vietnam, by 1968 they were moving between 10,000 and 20,000 troops a month into the country, along with food and ammunition. In all, perhaps 1,000,000 people traveled down the Ho Chi Minh Trail during the war.[37]

The Americans attempted to stop the flow along the trail. As early as 1961 the CIA launched a clandestine campaign in Laos, recruiting several thousand Hmong tribesmen to attack and sever the route. In addition, United States Army Special Forces operated advance camps near the trail outlets in the south, millions of leaflets were dropped along the route in psychological operations, and in 1971 a South Vietnamese Army invasion was launched into Laos, but failed. Ultimately, the Americans relied upon air attacks. Low level helicopter gunships proved the most effective against the trucks using the trail. Although they flew only a small percentage of the missions, they accounted for nearly half the destruction. But the helicopters were highly vulnerable, and so many were shot down by the several thousand antiaircraft artillerymen defending the trail that their use was suspended wholly or partly for much of the war. The bulk of the sorties, several hundred a day, were flown by tactical aircraft and B-52 bombers.[38]

The devastation the carpet bombing patterns could wreak was virtually apocalyptic in its totality, as Truong Nhu Tang's description testified. But as soon as each flight of B-52 "steel crows" finished bombing, work crews began repairing the damaged stretches of roadway while others constructed bypasses. If the trucks were stalled too long, human and bicycle porters would be brought in to transport goods in the interim. The net result was an increasingly complex web of main routes, bypasses and cutoffs that became ever more difficult to interdict. It was an amazing testimony to human persistence, commitment, and sacrifice as they used everything from hand labor to animals and bulldozers to keep it open. Strategically, the Ho Chi Minh Trail was "the only battle that really mattered—and the only one that never ceased."[39]

When Stanley Karnow flew over the trail in helicopters in the 1960s, he said that no road network was discernible, even at low altitudes, beneath the green canopy that seemed to stretch on endlessly. Likewise, in 1966 Sol Saunders, of *US News and World Report,* said that his "whole flight had an eerie quality. Although there was no doubt that we were flying over a heavily traveled road, I saw no sign of life during the entire time." By 1974 the Paris accords prevented the Americans from bombing, and there was no longer any need for camouflage. That August *The Bangkok Post* published South Vietnamese Air Force photos showing that the trail was, in places, a four-lane highway with asphalt surfaces. At approximately every 100 kilometers the route included what was effectively a village for the soldiers and work battalions that protected the lifeline, with smaller depots between. Tang drove north along the route in 1974. At night, at a constant speed of about 30 kilometers an hour and amidst a continual flow of traffic, a "beautiful spectacle would emerge before our eyes, an endless stream of flickering headlights tracing curved patterns against the blackened wilderness, as far in both directions as the eye could see."[40]

Homefront Hanoi

Although the bicycle has been a stalwart of civilian transport in urban Vietnam for many decades, in Saigon, residents radically increased their reliance upon motor vehicles, busses, and innumerable small motorcycles during the Second Indochina War. In Hanoi, in contrast, the few cars and jeeps were luxuries confined largely to VIPs. Aside from the jam-packed trams and walking, an estimated 300,000 bicycles were the basic transport.

When Harrison Salisbury visited Hanoi during Christmas, 1966, the American bombing was going on. He arrived at night and it was gloomy and difficult to see, the headlights of the occasional vehicles obscured, and the few street lights dimmed. The quiet, darkened city streets at first seemed deserted to him, until he gradually became aware of "vague files of bicycles silently moving in both directions." At first light he was amazed at the number pedaling past. As he moved about the city over the next few days, he observed not only thousands upon thousands of machines being ridden, but hundreds of bicycles parked outside one building or another. Many riders had leaves woven into their sun helmets for camouflage, which they wore night and day, and even some bicycles were camouflaged.

A new bicycle cost several months wages to buy, whether locally made or imported from China. The Hanoi cycle factory had planned to produce 100,000 bikes in 1965, but never reached its goal because of the war. With the need to keep men and materials moving, the Chinese imports became ever more critical. While a permit was needed to buy a cycle, they could be had on the black market at about five times the official price. One night Salisbury witnessed about 200 people standing quietly by two trucks, exchanging their old, worn out bicycles for newly manufactured ones. In such economic circumstances, among the more prized presents one could receive was a new bicycle chain or other part. A surprising fact to Salisbury was that in an otherwise "remarkably law-abiding" North Vietnam, bicycles were commonly stolen. He noted that "no one left a bike without a padlock on the wheel."[41]

In 1974, Michael Richardson, a *Sydney Morning Herald* reporter, visited Hanoi after Australia had established diplomatic relations. On the drive from the airport to the city, he saw innumerable bicycles and dozens of trucks, but almost no cars. One consequence was that Hanoi was virtually free of exhaust fumes and pleasantly quiet—there was only the "silent whoosh" of passing cycle traffic. It was also a field day for those interested in vintage bicycles. Of necessity, 50 old Rudge cycles, from Nottingham, England, were still being ridden. In contrast, when Australian diplomats set up shop in late 1973, they imported a number of "Easy-Rider" bicycles from Hong Kong for the staff's personal use. Their machines, with multiple gears, high handlebars, gold-flecked plastic seats, and padded backrests, stood out like sore thumbs.[42]

Thirty years of war took its toll upon the Hanoi populace. In 1974 Truong Nhu Tang returned to the city for the first time in three decades. He saw a vista of "crumbling facades and peeling paint, with only a hint of bygone dignity. The human landscape was equally wrenching. People walking or biking in the streets shared a look of grim preoccupation. Though the streets were crowded, there was none of the bustle or vitality that Asian cities usually display."[43]

The Tour de Propagande

The bicycle became something of a symbol in the propaganda battle that swirled around the war. In a gesture akin to shipping coals to Newcastle, the Polish people in 1967 emphasized their solidarity with their North Vietnamese brethren by sending them, along with 20 tons of condensed milk, 100 bicycles. The following year a French company, funded by a private organization, helped develop and build a special bicycle for North Vietnamese physicians and surgeons. It was smaller than traditional European models and contained surgical and medical kits and two headlights, with detachable extension cables, for lighting a small field hospital.[44]

In January, 1970 the South Vietnamese government held a bicycle race to show that it exercised "complete control over the countryside." Cycle racing, introduced by the French, was South Vietnam's leading sport and most of the country's Olympic representatives in 1968 were pedalers. In effect, the government revived an abbreviated version of the 1,360 mile Tour de Vietnam, which was suspended some 14 years earlier.

The six-day race left from Nha Trang, 190 miles north of Saigon, included a stretch through the central highlands, and ended in the Mekong Delta. The route passed Camranh Bay, but no United States soldier deigned to enter. One rider felt that since "it's a sporting event and the Vietcong know the people want to be able to watch it," they should be safe. A corporal from the South Vietnamese Rangers emphasized that "we're athletes, we have nothing to do with politics." Armed helicopters swarmed overhead, and relays of tanks and armored personnel carriers supported them en route, with a contingent of 150 Americans and Vietnamese sweeping the road ahead for mines. As expected, the Vietcong attacked the advance troops, killed at least one, and temporarily halted the race; their idea of sport was obviously different from that of the South Vietnamese.

Since none of the riders had been killed or injured, President Nguyen Van Thieu boasted that the race proved that "the government rules the road," and that one could drive safely from the Demilitarized Zone to the southern tip of the country. He neglected to mention that the race had started hundreds of miles south of the zone. Terence Smith, of *The New York Times*, concluded that the South Vietnamese government demonstrated only that it could conduct a bicycle race if several thousand soldiers were available to protect the riders. It was a case of sport and politics riding in tandem.

9. Retrospect and Prospect

There are reasons military bicycling never caught on, and this PBS documentary delineates them all.
　　　　　　KIMBERLY HEINRICHS, AMAZON.COM EDITORIAL REVIEW OF
　　　　　　THE BICYCLE CORPS: AMERICA'S BLACK ARMY ON WHEELS

With the onset of the new millennium, the Swiss Army announced that it would be dismantling its legendary cycling units, a step that had been prophesied since at least 1995. It received wide international coverage in newspapers, magazines and in broadcast programs, often with a smile. The BBC's Claire Doole reported in May, 2001, that the last remaining two-wheeled regiment would be disbanded by 2003, nearly 120 years after the machine's adoption by the Swiss military. A number of soldiers felt the decision short-sighted and lamented the passing of the units. However, it attracted none of the outcry that accompanied abolishment of the Swiss cavalry and pigeon carriers some thirty years before. In the view of some, that represented an unmistakable sign of the quaint machine's military obsolescence.[1]

　　However the bicycle is not obsolete. Many military services still employ large numbers of them, ranging from carrying them on patrol vehicles in Afghanistan, to aircraft technicians pedaling them about vast runway aprons to service aircraft. It is an innocuous device that, while not a weapon in its own right, can still be part of a deadly effective man-machine combination on the battlefield, or in support. Like the horse, jeep, and tank, it has particular characteristics, capabilities and limitations, and can only occupy certain niches. Each war also has its own set of niches, and many will not suit a bicycle—or a horse, jeep or tank. That does not mean that they are obsolete, only that they are momentarily irrelevant, a very different thing. If the history of military cycling teaches us anything, it is not only that obsolescence and usefulness lie in the eyes of the beholder, but that some see niches where others do not.

Swiss soldiers riding the older-style bicycles that were eventually replaced

By the 1920s certainly many believed the bicyle to be militarily obsolete, and history is replete with the records of numerous cycle units which had been formally established, trained, and

Swiss Army cyclist with the final model

equipped, which served no effective use, and were finally abandoned. Yet, the machine has continued to pop up in wartime ever since, in some unexpected places, and with sometimes astounding effect. Significantly, in many of those times and places the bicycle's users were not part of any formal cycle unit. They simply had a need, and the ubiquitous machine fulfilled it.

Among the most fascinating things about the bicycle's military use is that its importance has been ignored by some who should know better. And that tendency can be traced right back to its first significant use, in the Boer War. Major-General W. H. Mackinnon, who commanded the C.I.V. in South Africa, later wrote a journal about his unit. Anyone reading it, however, would never know that bicycles were used in that war, let alone that the largest single group of cycle troops—one thousand—were under his direct command. Likewise, a C.I.V. War Souvenir pamphlet had only a single reference to cyclists, and that in a caption. They are in stark contrast to the work of J. Barclay Lloyd, the C.I.V. soldier who left a detailed record of the bicycle's use in the war.[2]

Another striking example is Tadao Nakada's book, *Imperial Japanese Army and Navy Uniforms and Equipment* (in Japanese with an English language summary). It is a massive collection of illustrations of thousands of items relating to the Japanese involvement in World War II. It encompasses both official issue equipment and things accumulated by soldiers during their campaigns: uniforms, field equipment, weapons, posters, postcards, motorcycles, trucks, cars and so on. But despite the fact that the Japanese army made extensive and highly effective use of the bicycle, there is not a single illustration or reference to the machine in the book.[3]

In Peter Charlton's *War Against Japan 1941-1942*, written for the Time-Life series, *Australians at War*, forty per cent of the book's text, photos and accompanying captions are devoted to a detailed description of the invasion of Malaya and the fall of Singapore. However, the critical role played by the bicycle is the subject of only three trivial references. The only mention in the text was to the Australian ambush of several hundred Japanese at Gemencheh, who, as Charlton wrote, pedalled into it "laughing and chattering away." The other two references were in photograph captions. While Charlton specifically mentioned the Japanese tactics of "frontal fixing and local encirclement," he never referred to the integral role of bicyclists, and there is absolutely no indication that thousands of Japanese soldiers in fact used bicycles in the campaign.[4]

At a meeting of the U.S. Senate Foreign Relations Subcommittee on Africa, in January, 1976, the members were discussing the fact that the Ford administration had offered open financial aid to Angolan factions fighting Soviet-supported groups. In response to a pointed question on whether the United States might provoke large-scale Soviet intervention by authorizing the transmission of $3,000,000 to the National Front for the Liberation of Angola, the reply was that the money had only been used for bicycles and office equipment, not arms, how could anyone interpret those as warlike? The Vietnam conflict had ended less than two years before.[5]

Six years later, however, Barry Goldwater had no doubts as to the machine's nefarious side. As U. S congressional elections neared, the El Salvadoran civil war raged on and many debated whether the United States should continue playing a role there. As the guerrillas grew stronger, Secretary of State Alexander Haig

Flight-line technicians at Luke Air Force Base, Arizona

stated that they were receiving Soviet-sponsored support, based upon CIA information that he could not reveal. Given the Reagan administration's credibility problems over El Salvador, some of the source material was eventually released. Goldwater, then Chairman of the Senate's Select Committee on Intelligence, disclosed that as far as arms shipments from Cuba to El Salvador via Nicaragua were concerned, "We have definite proof from aerial photos and ground observations. They [arms] are put on small vessels, some of them small fishing boats. They are then taken up the coast, where they are met by other small fishing boats, and in some instances put on small aircraft and flown to the interior". He continued, "They are sometimes even put on bicycles and shipped inland to the rebels."[6]

The divergent views and passions aroused when considering the bicycle's wartime use was a fascinating element of this study. An Australian Colonel who served in Vietnam told me that he and others were routinely frustrated in their efforts to get those higher up in the command chain to appreciate how rapid and effective was the mobility and materiel support of Vietnamese traveling by bicycle and on foot. In contrast, another Australian Colonel's unequivocal view was that any importance or successes attributed to the Vietnamese use of bicycles was "propagandistic bullshit." The anger in his response echoed the 1888 account of Lord Wolseley having to deflect the "combative qualities" of Colonel Hale when he got into a heated debate with Lt. Colonel Savile over the merits of cyclists versus horses.

The anger extended beyond military circles. During the course of this study a number of civilian bicyclists indicated that they found it abhorrent that the anodyne machine has been dragged into the terrible and inhumane business of war and dealing death to others. Further, I was severely criticized by some of them for even researching and writing about the bicycle's wartime use. An attendant fact is that no general cycling magazine has accepted for publication any of my submitted material relating to military cycling, even though they have published various other cycling related articles of mine.

Learned articles about the future of man-powered military vehicles still occasionally appear, such as F. P. U. Croker's discussion of the future of man powered military vehicles in *The Army Quarterly and Defence Journal*. He reiterated the usual list of advantages (small, light, needs no fuel, relatively quick compared to foot soldiers, and able to use narrow routes not suited to larger vehicles), and indicated that the military bicycle "will retain a useful role during the foreseeable future". In 1975 Major Stephen Tate submitted to the Faculty of the U.S. Army Command and General Staff College, Fort Leavenworth Kansas, a Master of Military Art and Science Thesis titled 'Human Powered Vehicles in Support of Light Infantry Operations'. After surveying the history of bicycle use in wartime, detailed characteristics of the machine and the most recent technological advances, he concluded that with it light infantry units could 'significantly improve their practical mobility'.[7]

Neither added anything significant to what has been said before. However, that is not so much a criticism of them as realising that there is probably little new to say. The basic man-machine interaction that is "bicycling" has not substantially changed since the Boer War started; the power unit is still the same. Only the niches change.

J. F. C. Fuller wrote that "The object of military training is to prepare the soldier for the next war, for it is the only possible war in which he can fight". True, but past wars are the only ones in which soldiers have been able to learn from actual combat. So how easy is it to take even a simple military cycling lesson from the past and apply it effectively to the present? Not very, it would seem, as Rhodesia's Selous Scouts found out in 1976.[8]

During the Rhodesian government's campaign against FRELIMO guerrillas, a group of Selous Scouts wanted to confirm the location of a FRELIMO transit camp at Mapai, some 75 kilometres inside Mozambique, near the South African-Rhodesian border. In assessing the best route to get to Mapai and back, they decided to travel along a railway line in Mozambique. One of the officers then recalled a photograph of the railway war cycles in the Boer War. After detailed discussions with a Rhodesian railway engineer and confirmation of Mozambique's rail gauge, Lieutenant Colonel Ron Daly ordered the construction of a modern version of the two-man war cycle. He specified that it must be easy to strip and reassemble, as it would have to be hidden in the bush near the rail line while scouting in the adjacent areas. When finally delivered and shaken down in Rhodesia, it "worked like a dream and was silent", held enough food and water for 14 days in a nylon net slung between the two men, and could be pedalled easily at 20 - 30 kilometres per hour.

On the night of June 21, 1976, several Selous Scouts carried the bits and pieces of the cycle unit across the border to the Mozambique railway line, where they assembled it. Once it was placed on the track, however, they discovered to their dismay that the ballast along the line was so poor that not only the sleepers (ties) but the track itself had sunk into the ground in places. The line was unrideable. They dismantled the war cycle and returned to Rhodesia, on foot.[9]

Colonel Daly's men are not the first soldiers to find that what has happened in the past is no guide to the future. It is worth recalling that significant wartime uses of the bicycle—The Boer War, World War I; the German blitzkrieg; the Japanese roll down Malaya; or the Dien Bien Phu logistical effort—have each been markedly different from any previous major use.

Over the past quarter century there have been some clever new technologies and inventive efforts applied specifically to the development of folding military cycles. In 1987 Montague Bicycles was founded by an M.I.T. graduate student who invented the "Bi-Frame" folding bicycle, so-named for its patented Concentrus system, which nests one frame tube inside the other, providing an absolutely rigid unit when locked open. It overcame the instability problem confronting the long-term

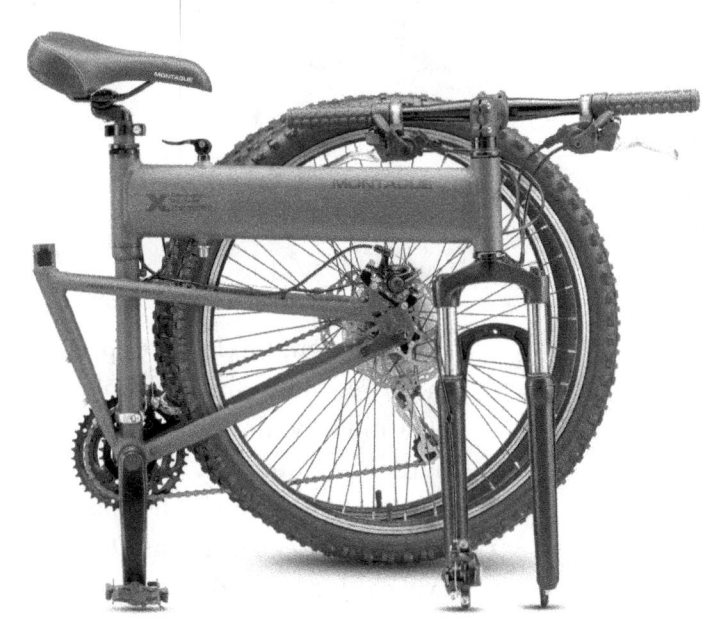

Montague 'Paratrooper' folding bicycle

users of most other folding bike designs and was soon being sold to the military. Appreciating the value of folding bikes, in 1997 the U.S. Defence Advanced Research Projects Agency awarded Montague a two-year grant involving testing and development of an electric mountain bike. However, after various engineering developments, and tests at Marine bases in Quantico, Virginia, Camp Pendleton, California, and Lejeune, North Carolina, both those overseeing the trials and the soldiers themselves concluded that the various technical problems and difficulties associated with the hybrid electric machine were simply not worth it, that they do not need the electric component. While the electric bike did not pan out, from this project came a new

US Marine on a Montague folding bicycle

frame design, the Paratrooper, which can be folded or unfolded in 30 seconds, with no tools, and fits through the cargo door of a military aircraft with an airborne ranger. Incorporating a hardened aluminum frame and wheels, it weighs only 29 lbs, but is extremely strong. The machine is now the world's standard military cycle, though, as David Montague observed, 'It is a very cyclical business, depending on military activity, who's got the budget, what's going on, and perceived needs'. [10]

The possibility of incorporating sophisticated electric motors in the bicycles came to momentary public prominence during the 2004 U.S. presidential campaign, when retired General Wesley Clark, former NATO supreme commander, was for a short while a potential candidate and the subject of widespread public discussion. General Clark had witnessed the value of bicycles in Vietnam and had become involved with the WaveCrest company, which hoped to interest police departments and the military. I was contacted by a *New York Times* writer at the time seeking my comment on the machine's viability. (The journalist said he had no idea bicycles had ever been used in wartime until he was assigned to write the article.) My immediate reaction was that while it might be good for police departments in urban areas, the motors and batteries not only brought the machine's weight to 64 pounds, but required charging and support facilities. I felt that those factors combined could negate much of the proven military

Folding bike mounted on a bomb disposal unit in Afghanistan

value of a bicycle in the field: its operational simplicity. Others interviewed for the article expressed similar and other reservations. The military did not adopt an electric version and the company went out of business in 2006, though other commercial interests have exploited the civilian possibilities, and several models are now available.[11]

As to the future, will the bicycle ever again see significant military use on a large scale? Who knows? While there are now computer simulators playing war games, they are only simulations, not crystal balls. Furthermore, future opponents may not program their own computers, or play their own games, with the same parameters or assumptions. They may not even use a computer at all.

And as for obsolescence, the use of military thermal technology has high-lighted a previously irrelevant, but now significant, element in the pantheon of bicycle characteristics: It gives off little or no thermal signature to belie its presence.

The power source still does, however.

Notes

Chapter 1 (The Experimental Years)

1. The concept of "bicycling" is developed in great detail in Jim Fitzpatrick, "The Bicycle in Rural Australia: A Study of Man, Machine and Milieu," a Ph.D. dissertation completed at the Australian National University in 1979, and published in 1980 as *The Bicycle and the Bush: Man and Machine in Rural Australia.*

2. Innumerable books and articles on bicycling technology and history have been produced and more arrive on a regular basis, but this is not the place to review the field. David Herlihy's book, *Bicycle: The History*, is an excellent introduction and good place to start, as is David Gordon Wilson's *Bicycling Science*, 3rd edition, the definitive work of its kind. The various Proceedings of the International Cycling History Conference (www.cyclinghistory.org) give some idea of the variety of research interests.

3. David Gordon Wilson, "Human Muscle Power in History," in James C. McCullagh, ed., *Pedal Power*, pp. 1-36; H. O. Duncan, *The World on Wheels*, vol. 1, p. 278; "Bicycles Long Used for Military Service," *The Bicycling World and Motorcycle Review*, October 6, 1941, p. 36; "The Military Bicycle," *Scientific American*, Nov. 30, 1889, p. 343.

4. This material was condensed from various sections of Fitzpatrick, *The Bicycle and the Bush*. It specifically cites information from Norris McWhirter, *Guinness Book of Records;* Whitt and Wilson, *Bicycling Science*, p. 194; Caunter, *The History and Development of Cycles*, pp. 37, 40, 47-48; A. C. Pemberton, *The Complete Cyclist*, p. 209.

5. J. Barrow, "Departmental Bicycles-Average Cost of Repair," report submitted to the Chief Inspector, Stores and Transport, Postmaster-General's Department, Melbourne, by the Superintendent of Stores and Transport, August 6, 1940; L. Hammond, Report submitted to the Superintendent of Stores and Transport, Postmaster-General's Department, Melbourne, by the Senior Cycle Mechanic, July 26, 1940; Postmaster-General (Australia), "Departmenal Bicycles-Renewal of," memo sent from the Chief Inspector, Stores and Transport, Postmaster-General's Department, Melbourne, to the Director-General, Melbourne, May 21, 1935; "Memo" sent from the Chief Inspector, Stores and Transport, Postmaster-General's Department, Melbourne, to the Deputy-Director, Melbourne, November 27, 1935; "Departmental Bicycles-Average Cost of Repair," letter sent from the Chief Inspector, Stores and Transport, Postmaster-General's Department, Melbourne, to the Superintendent, Stores and Transport, all states, April 24, 1940; P. A.

Weeks, "Departmental Bicycles-Average Cost of Repairs," report submitted to the Chief Inspector, Stores and Transport, Postmaster-General's Department, Melbourne, July 16, 1940.

6. A .E. Harrison, "The Competitiveness of the British Cycle Industry," *Economic History Review*, vol. 22, no. 2, 1969, pp. 287-303; *The Austral Wheel*, Nov. 1896, p. 319.

7. This material was condensed from Fitzpatrick, *The Bicycle and the Bush*, "Environment and Adaptation," pp. 96-149. It drew upon a wide array of personal interviews and published and unpublished sources relating to bicycle tires and cycling in various conditions, for the period 1890-1930.

8. *The Austral Wheel*, August 1896, p. 231.

9. Lieutenant Henry H. Whitney, "The Adaptation of the Bicycle to Military Uses," pp. 543-548; "Army Cyclists," *Journal of the Society for Army Historical Research*, vol. 38, No. 153, March 1960, p. 43; Henry Sturmey, ed., *The Cyclist Year Book 1898*, pp. 100-106; Ernest J. Martin, "The Cyclist Battalions and their Badges, 1888-1921," *Journal of the Society for Army Historical Research*, vol. 22, no. 91, Autumn 1944, p. 277.

10. Eamon O'Dea, "When Connecticut Put the Cavalry on Bicycles," *Yankee*, August 1979, pp. 68-71; Smith, *A Social History of the Bicycle*, pp. 227-228.

11. "Military Cycling," *The Saturday Review*, June 16, 1888, p. 722.

12. Pemberton, "The Complete Cyclist," p. 352; "The Cycle and the Service," *The Sketch Cycling Supplement, Stanley Show*, September 30, 1896, p. 408; J. R. Williams, "Push Bike Soldiers," *Firm & Forester*, vol. 5, no. 2, November, 1978, p. 125; Whitney, "The Adaptation of the Bicycle," pp. 553-554.

13. The Earl of Albemarle and G. Lacy Hillier, *Cycling*, pp. 102-104.

14. *The Austral Wheel*, March 1896, p. 72; Charles Turner, "Military Cycling," *Outing*, p. 192; First Lieutenant R. G. Hill, "The Capabilities and Limitations of the Bicycle as a Military Machine," *Journal of the U. S. Military Service Institutions*, July-Nov. 1895, p. 313. Hill's work is one of the most thoughtful written in the 1890s. Percy Armstrong, "Cyclists for the Transvaal," *The Western Mail*, Feb. 17, 1900, pp. 18-19.

15. Whitney, "The Adaptation of the Bicycle," p. 554.

16. "French Wheelmen," *Army and Navy Journal*, Nov. 14, 1896, p. 182; Charles Turner, "Military Cycling," *Outing*, p. 192.

17. "Army Bicycles," *Scientific American*, August 18, 1894, p. 100; Whitney, "The Adaptation of the Bicycle," p. 555; O'Dea, "When Connecticut Put the Cavalry on Bicycles," pp. 68-69; "Military Bicycling," *Scientific American*, September 19, 1891, p. 185; untitled article on military cycling, *The Chautauquan*, Feb. 1897, p. 568; Major Howard A. Giddings, "The Bicycle in the Army," *Harper's Weekly*, p. 364; *The Austral Wheel*, April 1896, p. 89.

18. Whitney, "The Adaptation of the Bicycle," pp. 551.

19. "The Bicycle Relay Race Across the Continent," *Scientific American*, Sep-

tember 19, 1896, p. 238; "The Cycle and the Service," *The Sketch Cycling Supplement, Stanley Show,* Sept. 30, 1896, p.407.

20. Lieutenant De R. C. Cabell, "Military Bicycling Thro' the Dakotas," *Outing,* June 1896, p. 216; Whitney, "The Adaptation of the Bicycle," pp. 551-552, 554; Hill, "The Capabilities of the Bicycle," *Journal of the U.S. Military Service Institutions,* July-Nov. 1895, p. 314; Giddings, "The Bicycle in the Army," p. 364; Lieutenant Colonel F. Maurice, *Sir Frederick Maurice: A Record of his Work and Opinions, with Eight Essays on Discipline and National Efficiency,* p. 100.

21. Whitney, "The Adaptation of the Bicycle," p. 557; Hill, "The Capabilities of the Bicycle," p. 313, 315; Giddings, "The Bicycle in the Army," p. 364.

22. Whitney, "The Adaptation of the Bicycle to Military Uses," p. 556; "The Easter Cyclist Manoeuvres," *The Saturday Review,* April 7, 1888, p. 411; Lieutenant William T. May, "On the March," 1892, p. 33; "The War Bicycle," *Scientific American,* March 23, 1897, p. 187; Hill, "The Capabilities of the Bicycle," p. 321.

23. Hill, "The Capabilities of the Bicycle," p. 315.

24. Cabell, "Military Bicycling," pp. 214-220.

25. The following information, unless otherwise noted, is taken from Lieutenant James A. Moss, "Bicycle Corps, 25th Inf.," *Army and Navy Journal,* Sept. 5, 1896, p. 11; "The Bicycle for Military Purposes," *Army and Navy Register,* Aug. 29, 1896, pp. 130-131; "Recent Experiments in Infantry Bicycling Corps," *Outing,* Feb. 1897, pp. 488-492.

26. The following is taken from the series of articles written by Moss, "25th Infantry Bicycle Corps," *Army and Navy Journal,* July 3, 1897, p. 814; July 31, 1897, p. 807; Aug. 7, 1897, p. 903; Oct. 2, 1897, pp. 71-72; and "Experimental Bicycle Trip in the United States Army," *Scientific American,* June 5, 1897, p. 354. An example of overseas reporting was "The Bicycle in War," *The Austral Wheel,* June 1897, p. 188.

27. Moss, "25th Infantry Bicycle Corps," Oct. 2, 1897, p. 71.

28. File No. 70545, Record Group 94, National Archives, cited in Charles M. Dollar, "Putting the Army on Wheels: The Story of the Twenty-Fifth Infantry Bicycle Corps," *Prologue,* Journal of the National Archives, vol. 17, no. 1, Spring 1985, pp. 6-23. Dollar has researched original materials from the records of the Adjutant General's Office, National Archives, that add some valuable observations relating to the bureaucratic aspects of the rides, and the extent of support Moss received from General Miles. As well, he noted some interesting comments on how the cyclists were viewed by those in the towns they passed through. Given the excellent photographs included in the article, and his summary of the general cycling craze of the era, it is in several respects the best article on the ride I have seen.

29. Major R. P. Davidson, "Some Recent Experiments in Military Bicycling," *The Army and Navy Journal,* Aug. 7, 1897, p. 902.

30. Exhibit, National Army Museum, London, personal visit, April, 1983; A. S. Clark, ed., *The London Cyclist Battalion: A Chronicle of Events Connected with the 26th Middlesex (Cyclist) V.R.C., and the 25th (C. ofL.) Cyclist Battalion, The London Regiment, and Military Cycling in General*, p. 10; Major F. W. A. Hobart, *Pictorial History of the Machine Gun*, picture 32.

31. "Bicycle Artillery," *Scientific American*, Dec. 28, 1901, p. 21735.

32. Smith, *A Social History of the Bicycle*, pp. 231, 239; "New Army Bicycles," *Scientific American*, Feb. 8, 1896, p. 91. *The Austral Wheel*, Feb. 1896, pp. 300-301.

33. "The Military Bicycle, *Scientific American*, Nov. 30, 1889, p. 343.

34. Clark, ed., *The London Cyclist Battalion*, pp. 84, 270.

35. *La Nature's*, article was reprinted as "Bicycle Artillery," *Scientific American*, Dec. 28, 1901, p. 21735; G. Lacy Hillier, "The Use of the Cycle for Military Purposes," *Longman's Magazine*, July 1887, p. 272.

36. "The Folding Military Bicycle," *Scientific American Supplement, No. 1044*, Jan. 4, 1896, p. 16681.

37. H. Gerard, "Le Probleme de I'Infanterie montee resolu par I'emploi de la bicyclette," Paris, 1894; "The Folding Military Bicycle," *Scientific American Supplement, no. 1044*, Jan. 4, 1896, pp. 16681-16682; "The Bicycle in the Army," *Scientific American Supplement, no. 1086*, October 24, 1896, p. 17358; Major John M. Macartney, "Portable Bicycle for Mobile Infantry," *The United Service Magazine*, vol. 14, (n.d.) pp. 388-395.

38. "French Wheelmen," *Army and Navy Journal*, Nov. 14, 1896, p. 182; "The Folding Military Bicycle," *Scientific American Supplement, no. 1220*, May 20, 1899, p. 19559; "French Military Folding Bicycle," *Scientific American*, Oct. 26, 1901, p. 264; H. Gerard, "Infanterie Cycliste in Campagne".

39. "A Folding Bicycle," *Scientific American*, March 13, 1897, p. 165; "Bicycling in the Austrian Army," *Scientific American Supplement*, no. 1121, June 26, 1897, pp. 17911-17912.

40. "A New Military Folding Bicycle," *Scientific American*, Oct. 20, 1900, p. 244.

41. "The Bicycle in the Army," *Scientific American Supplement*, no. 1086, Oct. 24, 1896, p. 17358; *L'illustration's* article was reprinted as "The Bicycle in the Army," *Scientific American Supplement*, no. 1086, Oct. 24, 1896, p. 17358.

42. Whitney, "The Adaptation of the Bicycle," pp. 559-561.

43. Fitzpatrick, "The Early Development of Australian Road Maps," *The Globe, Journal of the Australian Map Curators' Circle*, no. 13, pp. 13-29.

44. *The Australian Cyclist*, Feb. 20, 1896, p. 5; Whitney, "The Adaptation of the Bicycle," p. 552; "Cycle Telegraphs in War," *Scientific American*, Aug. 29, 1896, p. 187.

45. The following summary is from various portions of Wilfred Pollock's *War and A Wheel: The Graeco-Turkish War as Seen from a Bicycle*. The book is interesting reading not only in terms of cycle use, but for its descriptions of the general method of reporting wars at the time.

46. "Bicycling in the Austrian Army," *Scientific American Supplement*, No. 1121, June 26, 1897, pp. 17911-17912.

47. Whitney, "The Adaptation of the Bicycle," pp. 553, 557; Arthur Judson Palmer, *Riding High*, p. 148; "Bicycling in the Austrian Army," *Scientific American*, no. 1121, June 26, 1897, pp. 17911-17912.

48. "Military Bicycling," *Scientific American*, Sept. 19, 1891, p. 185; "A Manual of Cyclist Drill," *The Saturday Review*, Aug. 2, 1890, p. 18; Whitney, "The Adaptation of the Bicycle," p. 555; "The Military Bicycle," *Scientific American*, Nov. 30, 1889, p. 343.

49. *The Morning Herald* (Western Australia), Aug. 31, 1897, p. 2; "25th Infantry Bicycle Corps," *Army and Navy Journal*, July 3, 1897, p. 814.

50. "Military Cycling," *The Australian Cyclist*, Feb. 6, 1896, p. 9.

51. Whitney, "The Adaptation of the Bicycle to Military Uses," pp. 542, 563; Hill, "The Capabilities of the Bicycle," p. 322.

52. *Cycling Supplement*, Stanley Show, Sept. 30, 1896, p. 408.

53. Henry Sturmey, *The Cyclist Yearbook* 1898, pp. 100-106; A. C. Pemberton, *The Complete Cyclist*, p. 44; "The War Office and Cycles," *The Cyclist*, Aug. 3, 1898, p. 859.

54. Whitney, "The Adaptation of the Bicycle," pp. 542, 563; Giddings, "The Bicycle in the Army," p. 364, duplicates much of Whitney's information, but provides some supplementary data which I have incorporated.

55. Giddings, "The Bicycle in the Army," p. 364, also cited Italy as having had cyclist soldiers in 1870.

56. Macartney, "Portable Bicycle for Mobile Infantry," p. 390.

Chapter 2 (The Boer War)

1. My basic reference was Thomas Pakenham's *The Boer War*, considered the definitive study. He overlooked the use of the bicycle but, between my correspondence and a personal visit and long discussion about the war and cycle use, assisted me with valuable leads. Rayne Kruger's *Good-bye Dolly Gray: The Story of the Boer War*, provided further background.

D. R. Maree's "Bicycles in the Anglo-Boer War of 1899-1902," *Military History Journal*, vol. 4, no. 1, June 1977, pp. 15-21, was later published separately by the South African National Museum of Military History as *Bicycles During the Boer War*, 1899-1902. Maree assembled information from a wide array of materials. Unfortunately, several of the cited references do not support or correspond with his text, which made follow-up work difficult. Nonetheless, it is a significant contribution to the field.

Special mention is required of H. W. Wilson's work, *With the Flag to Pretoria: A History of the Boer War of 1899-1900*, 2 vols.; and *After Pretoria: The Guerrilla War. The Supplement to With the Flag to Pretoria*, 2 vols. They provided the best sources of photographs that I found, with respect to both the Boer War in general and the nature and extent of cycle use in particular. However, Wilson had the frustrating habit of including some remarkable photographs of wartime cycle use with little or no accompanying comment, either in text or caption.

2. *The Standard and Diggers'News*, Jan. 18, 1896, pp. 4, 11; and H. M. Hole, *The Jameson Raid*, p. 188, cited in Pakenham, *The Boer War*, p. 5.

3. Elizabeth Pakenham, *Jameson's Raid*, pp. 86-89, referred to the messenger cyclists, as did *The Cape Times* on Jan. 2, 1896, p. 5; Jan. 3, 1896, p. 7; Jan. 9, 1896, p. 5; and in "Crack Cyclist Arrested," Jan. 7, 1896, p. 5. The Osmond bicycle advertisement appeared in *The Austral Wheel*, Aug. 1898, p. 221, among others.

4. The estimate of numbers of bicycles on the goldfields is from J. R. Shorten, *The Johannesburg Saga*, p. 174, cited by Maree, *Bicycles in the Anglo-Boer War of 1899-1902*. English cycle sales figures for 1897 are in Henry Sturmey, ed., *The Cyclist Yearbook 1898*, p. 142. Reference to legislation is in *The Cape Times Weekly Edition*, Dec. 8, 1897, p. 5.

5. Joseph Chamberlain, to Sir Michael Hicks Beach, the Chancellor of the Exchequer, quoted in Pakenham, *The Boer War*, p. 84.

6. Thomas Pakenham, *The Boer War*, pp. 288, 307. Troop figures are from Pakenham, p. 572. The observation balloon was described by Kruger, *Good-bye Dolly Gray*, pp. 243-244.

7. The blockhouse system's design and expansion are discussed in detail by Kruger, *Good-bye Dolly Gray*, pp. 428-429, 441; Pakenham, *The Boer War*, pp. 536-7, 546-547; and in Martin H. Brice's *Stronghold: A History of Military Architecture*, pp. 158-159. Communications along the system are discussed in R. F. H. Nalder, *The Royal Corps of Signals: A History of Its Antecedents and Development*. The stated number of blockhouses built and miles of corridors vary from source to source.

8. The detailed description of the war cycle's construction and first trials appeared in "A Railway Cycle," *Official Journal of the Cape Town Cycling and Athletics Club*, New Series, no. 12, March 1900, pp. 7-8. His name appeared as "Menzas" in another brief reference, but "Menzies" was used in the longer and more detailed reports.

9. *Black and White Budget*, March 31, 1900, p. 11.

10. Wilson, *With the Flag to Pretoria*, vol. 2, p. 914; *The Cape Times Weekly Edition*, July 30, 1902, p. 8.

11. Information provided by Mr. Andy May, Fort Klapperkop, via General Philip Pretorius, Director, South African National Museum of Military History, personal correspondence, Jan. 20, 1993.

12. See, for example, the reprinted 1908 *Sears, Roebuck Catalogue*, p. 173.

13. Theron and the Boer struggles are discussed extensively in Michael Davitt, *The Boer Fight for Freedom*; and Frederick Rompel, *Heroes of the Boer War*. Lord Roberts' comment is from Maree, *Bicycles in the Anglo-Boer War of 1899-1902*, citing the Roos Telegrams, A 285, Transvaal Archives.

14. For reference to various patrols see L. S. Amery, *The Times History of the War in South Africa*, vol. 2, pp. 84-85; vol. 3, p. 81; Major-General Sir Frederick Maurice, *History of the War in South Africa*, vol. 2, pp. 17, 106.

15. Maurice, p. 263.

16. Major E. A. Altham, "Notes on the Lines of Communications in the Cape Colony," War Office: Intelligence Division, June, 1899; *The Cape Cyclist*, Oct. 1899, p. 10; *Black and White Budget*, June 16, 1900, p. 326; Johannes Meintjies, *The Anglo-Boer War 1899-1902: A Pictorial History*, photo 88; Wilson, *With the Flag to Pretoria: A History of the Boer War of 1899-1900*, vol. 2, p. 603.

17. J. Barclay Lloyd, *One Thousand Miles with the C. I. V*, is floridly written and in its reading reflects the tedium of the war. However, it provides detailed insights into cycle usage. The examples cited in the following paragraphs are only a portion of his comments.

18. Fitzpatrick, *The Bicycle and the Bush*, pp. 138-141, 146-148.

19. *The Royal Commission on the War in South Africa*, Minutes of Evidence, vol. 2, p. 483.

20. "Boer War," *The Australian Encyclopaedia*, vol. 2, p. 45.

21. Fitzpatrick, *The Bicycle and the Bush*, pp. 158-166.

22. Percy Armstrong, "Cyclists for the Transvaal," *The Western Mail*, Feb. 17, 1900, pp. 18-19.

23. General background on Australians in the Boer War is from R. L. Wallace, *The Australians at the Boer War*, and L. M. Field, *The Forgotten War: Australian Involvement in the South African Conflict of 1899-1902*. The Queensland reference is from data provided by R. L. Wallace, personal correspondence, April 26, 1981.

24. Fitzpatrick, "Arthur Richardson," *Australian Dictionary of Biography*, vol. 11, p. 379; Fitzpatrick, *The Bicycle and the Bush*, pp. 174-178; Lieutenant Colonel P. L. Murray, ed., *Official Records of the Australian Military Contingents to the War in South Africa*, pp. 407-412; Kruger, *Good-bye Dolly Gray*, p. 65.

25. Kruger, *Good-bye Dolly Gray*, pp. 287, 297; Amery, *The Times History of the War in South Africa*, vol. 4, pp. 202, 207, 538.

26. Lieutenant Wynyard Joss, personal letter to Mr. C. Hopkins, May 29, 1901, made available courtesy of R. L. Wallace.

27. "Permits for Cyclists," *The Standard and Diggers' News*, October 14, 1899, p. 5; H. W. Wilson, *After Pretoria*, vol. 1, p. 493.

28. "British Spies on Bicycles," *The Standard and Diggers' News*, Oct. 21, 1899, p. 3; and Oct. 27, 1899, p. 4. Maree, in *Bicycles in the Anglo-Boer War of 1899-1902*, cited "The Siege of Kimberley," *The Diamond Fields Advertiser*, (n.d).

29. *The Cape Times Weekly Edition*, Feb. 6, 1901, p. 8; Wilson, *After Pretoria*, vol. 1, pp. 276-277; *The Cape Times Weekly Edition*, Feb. 6, 1901, p. 13.

30. Whitney, "The Adaptation of the Bicycle," pp. 558; B. A. H. Parritt, *The Intelligencers: The Story of British Military Intelligence Up to 1914*, pp. 210-211.

31. My estimate is based upon viewing and reading many photographic and written sources. See, for example, Commander C. N. Robinson, *The Transvaal War Album-The British Forces in South Africa*, pp. 98 and 132; and *The Cape Times Weekly Edition*, Oct. 25, 1899, p. 8. Maurice's estimate, from the *History of the War in South Africa*, vol. 3, was cited by Maree, *Bicycles in the Anglo-Boer War of 1899-1902*.

32. Kitchener's appeal was cited in Amery, *The Times History of the War in South Africa*, vol. 6, p. 274; the Cape Colony Cycle Corps formation in Amery, *The Times History of the War in South Africa*, vol. 5, p. 612.

Chapter 3 (The War of the Words: 1900-1914)

1. O'Dea, "When Connecticut Put the Cavalry on Bicycles," p. 71.

2. *The Austral Wheel*, Oct. 1899, p. 269; Clark, ed., pp. 84, 270.

3. Major B. G. Simpkins, *A History of the Rand Light Infantry*, pp. 1-7. 12-13; H. W. Wilson, *With the Flag to Pretoria*, vol. 2, p. 970.

4. Field Marshal Viscount Kitchener of Khartoum, *Memorandum on the Defence of Australia*, Melbourne: Government Printer, 1910, pp. 3-7. Colonel W. T. Bridges, "Report by Col. W. T. Bridges on Administration and Functions of the General Staff," CA 6, Australian Department of Defence [I], Commonwealth Record Series A289, 1909; Chief of the General Staff (Australia), "Training of Regimental Signallers," Defence files 1984/1/43; and Australian Archives, CRS B 414, entry 1849/1/25, and CRS B 414, entry 413/6/11. Personal communication with J. Vatts, Acting Senior Archivist, Australian Archives, Brighton, Sept. 16, 1980, confirmed that the Registers of Inwards Correspondence indicate that there were numerous references to the use of bicycles within the Defence Department during the period under consideration, but that most of them have not been transferred to the Archives, presumed destroyed. This is similar to the situation I encountered in attempting to research Australian Post Office records. Those responsible for culling records prior to transfer to the Archives presumably assumed that bicycles would be of no interest to posterity.

5. The 1909 and 1912 Dunlop relay rides are described in detail by Fitzpatrick in *The Bicycle and the Bush*, pp. 214-219, and the Coorong map in his article "The Early Development of Australian Road Maps," *The Globe, Journal of the Australian Map Curators' Circle*, no. 13, pp. 13-29. See also Harry James, "Looking Back," *Dunlop Gazette*, Aug. 1939, pp. 6-7, 12-13, and *The Argus* (Melbourne), Jan. 16, 1909, concerning the Coorong; W. J. Proctor, letter to Captain R. H. M. Collins, C. M. G., Secretary for Defence, Victoria Barracks, Melbourne, Jan. 20, 1909; S. A. Pethebridge, letter from the Acting Secretary, Department of Defence, to Dunlop Rubber Co. Ltd., Feb. 25, 1909.

6. *The Times*, April 12, 1909, p. 5; Colonel William T. Bridges, letter to the Secretary, the Dunlop Rubber Company of Australasia, Ltd., May 17, 1909.

7. *The Daily Telegraph* was partly owned by Hugh McIntosh, a publicist par excellence, and among the world's best sports promoters. He was responsible for staging the first black-white heavyweight boxing match, between Jack Johnson and Tommy Burns, in 1908, at Sydney's Rushcutters Bay. Before that, he had brought the world's most famous and highest paid athlete of the era, black racing cyclist "Major" Taylor, to Australia to race in the world's richest cycle race, which McIntosh also created and sponsored. Taylor's Australian tour was the subject of the 1992 international television mini-series, *Tracks of Glory*,

based upon the book manuscript by Fitzpatrick, *Major Taylor in Australia*.

8. Lieutenant David Beake, "Report on Dunlop Relay Ride. Easter 1912," A & I Staff, Australian Archives, 1912; Lieutenant Edwin Nixon, "Report on Military Aspect of Dunlop Relay Ride from Adelaide to Sydney, Carried Out During Easter 1912," A & I Staff, May 13, 1912; *The Australasian*, April 17, 1909, pp. 960-961.

9. Australian Imperial Force, *Notes on the Belgian Army*, Melbourne: Government Printer (n.d.) [post-1906].

10. Lieutenant J. Cook Wilson's *A Manual of Cyclist Drill for the Use of the Cyclist Section of the Oxford University Rifle Volunteer Corps*, was reviewed as "A Manual of Cyclist Drill," *The Saturday Review*, Jan. 4, 1890, p. 18.

11. See, for example, War Office, *Trumpet and Bugle Sounds for the Army*, 1914; *Instruction for Armourers . . . and for the Care of Bicycles*, 1904; *Instruction for the Care and Preservation of Bicycles*, 1904 and 1905; Major J. H. V. Crowe, *Handbook of the Armies of Sweden and Norway*, London: HMSO, 1901; General Staff, *Handbook of the Swiss Army*, London: HMSO, 1911.

12. General Albert Ordway, *Cycle-Infantry Drill Regulations of the District of Columbia National Guard*, Lieutenant William T. May, *Cyclists' Drill Regulations, United States Army*; Captain Howard A. Giddings, *Manual for Cyclists for the Use of the Regular Army, Organized Militia, and Volunteer Troops of the United States*, 1898.

13. A sample of the various of European army cycling manuals prepared at the time are those of the Ministere del la Guerre (Belgium), *Instruction pour les Compagnies Cyclistes*, 1906, and the Krigsministeriet (Denmark) publications, *Instruktion for Cyclens Brug i Haeren*, 1900, and *Cykelreglement for Haeren*, 1910.

14. Major-General Sir Frederick Maurice, "Cycling and Home Defence," *The Times*, May 24, 1901, p. 3; Arthur Conan Doyle, *The Great Boer War*, pp. 518-519; Captain A. H. Trapman, "The Cycle in Warfare: Its Potency as a Strategical and Tactical Factor," *The Times*, Dec, 17, 1908, p. 9.

15. "War on Wheels," *Black and White Budget*, Aug. 18, 1900, p. 638.

16. Walter Richards, *His Majesty's Territorial Army*, vol. 3, 1909, cited in Seamus McGonagle, *The Bicycle in Life, Love, War and Literature*, p. 103.

17. An illustration of a cyclist soldier/messenger in the Russo-Japanese War appeared in *The Town & Country Journal*, March 2, 1904, p. 19. The extent to which bicycles were used in that war is one of the most intriguing questions of this study. Several historians alluded to it, but could not recall reference details. I trust that someone with research knowledge of the war, and access to Japanese and Russian materials, may eventually be able to throw more light upon it. The Balkan and Tripoli references were from "The Beginnings of War-Machines: Military Bicycles," *The Illustrated War News*, Feb. 21, 1917, p. 11; it was one of many such retrospective military cycling references to come out of World War I.

18. B. H. Liddell Hart, *The Tanks: The History of the Royal Tank Regiment*, pp. 13-15; H. G. Wells, "The Land Ironclads," *The Strand Magazine*, 1903, reprinted in 1916 (my citations from *The War in the Air and Other War Forebodings*, New

York: Charles Scribner's Sons, 1926, pp. 383-415); T. H. E. Travers, "Future Warfare: HG Wells and British Military Theory, 1895-1916," in Brian Bond and Ian Roy, eds., *War and Society: A Yearbook of Military History*, London: Croom Helm, pp.67-87.

19. H. G. Wells, "The Cyclist Soldier," *Fortnightly Review*, vol. 74, Dec. 1900, pp.914-928.

20. Lieutenant Colonel Eustace Balfour, "Military Cycling, After Mr. H. G. Wells," *Fortnightly Review*, vol. 75, Feb. 1901, pp. 294-303.

21. Wells, "The Soldier Cyclist," *Fortnightly Review*, vol. 75, 1901, pp. 572-574.

Chapter 4 (World War I: The Allied View)

1. Daniel Yergin, *The Prize, the Epic Quest for Oil, Money, and Power*, pp. 167, 171-172.

2. John Ellis, *Eye-Deep in Hell*.

3. Barbara Tuchman, *The Guns of August*, pp. 49, 54.

4. The official British figures are from *Statistics of Military Effort of the British Empire in the Great War*, p. 173, cited by Lieutenant Ernest J. Martin in "The Cyclist Battalions and Their Badges, 1888-1921," in *Journal of the Society for Army Historical Research*, vol. XXII, no. 91, Autumn 1944, pp. 277-280; French and Belgian estimates are from Captain R. L. K. Allen, "Cyclist Units," *Journal of the Royal United Service Institution*, vol. 80, 1935, pp. 110-111.

5. *Manual of Military Law*, War Office, London, pp. 246, 297, 370-375; *The Times* (London), Aug. 19, 1916, p. 9.

6. Lieutenant E. W. Shaul, "A Memory of the Advance to the Aisne," *The Britannia*, no. 8, Oct. 1930, pp. 51-53; "English Cycle Rider Saves French Column," *The Bicycling World and Motorcycle Review*, Oct. 6, 1914, p. 36; "With the Cyclists in the Fields of Battle: An English Correspondent Narrates Some Exciting Experiences With and Between the Opposing Armies," Oct. 6, 1914, p. 37; "Despatch-Riders Give Thrilling Accounts of Their Adventures Mid Shot and Shell," Jan. 19, 1915, pp. 27-28; "From Paris to Liege," *Cycling*, Oct. 5, 1916, pp. 276-278; *Cycling*, March 8, 1917, p. 184.

7. "Cyclists' Work in Modern Warfare," *Cycling*, May 23, 1918, pp. 392-393; "A French Cyclist in the Battle of the Marne and the German Retreat," *Cycling*, Sept. 27, 1917, pp. 226-228, and Oct. 4, 1917, pp. 238-239; "With the Cyclists in the Fields of Battle: An English Correspondent Narrates Some Exciting Experiences with and Between the Opposing Armies," *The Bicycling World and Motorcycle Review*, Oct. 6, 1914, p. 37; G. W. Holderness, "Cyclists in Salonika: Life on Struma Front," *The Mosquito*, no. 129, March 1960, pp. 3-7.

8. "From Liege to Paris: The Story of a French Cyclist," *Cycling*, Feb. 21, 1918, pp. 150-153.

9. John Laffin, *Damn the Dardanelles*, 1980, p. 56; *Cycling*, Dec. 12, 1918, p. 392; Fairfax Downey, "It Wasn't Always Boxcars," *The American Legion Weekly*, Sept. 18, 1925, pp. 10-11, 20-21.

10. *The Illustrated War News*, April 28, 1915, p. 44; Clark, ed., pp. 95-124; "Indian Cyclists in France," *Cycling*, Aug. 3, 1916, p. 90.

11. The material for this section came from several sources: The personal diary of Major Jack Hindhaugh, made available through the courtesy of his daughter, the late Cecily Adams; "By Officers," *The Regimental History of New Zealand Cyclist Corps in The Great War 1914-1918*, 1922; Various 1st Anzac Cyclist Battalion records at the Australian War Memorial, listed in the "Sources" under "Australian War Diaries" and "Australian War Memorial"; and articles and papers by Fitzpatrick, including "Anzacs at War on Bicycles," *Royal Historical Society of Victoria Journal*, vol. 54, no. 3, 1983, pp. 31-38; "The Bicycle and the Australian Military—1890-1918: A Study of the Perception and Use of Technology," *Second Military History Conference*, Australian Memorial, Canberra, 1982; "The Bicycle and the Australian Military: 1890-1918," *Hemisphere*, vol. 27, no. 6, May/June 1983, pp. 341-346; and "War on Wheels: Major Jack Hindhaugh, Commander of the 1st Anzac Cyclist Battalion," *This Australia*, vol. 2, no. 1, Summer 1982-83, pp.34-38.

12. *Cycling*, Nov. 30, 1916, p. 446. See, for example, *The Illustrated War News*, May 26, 1915, p. 37; July 28, 1915, p. 10-11; Sept. 8, 1915, p. 11; Sept. 29, 1915, p. 25.

13. See, for example, *The Illustrated War News*, May 26, 1915, p. 37; July 28, 1915, pp. 10-11; Sept. 8, 1915, p. 11; Sept. 29, 1915, p. 25.

14. "Italy's Crack Cyclist Corps," *Cycling*, March 15, 1917, pp. 208-210.

15. John Terraine, "The Spectre of the Bomber," *History Today*, vol. 32, April 1982, pp. 4-9; *Cycling*, July 26, 1917, p. 63.

16. "Despatch-Riders Give Thrilling Accounts of Their Adventures Mid Shot and Shell," *The Bicycling World and Motorcycle Review*, Jan. 19, 1915, pp. 27-28.

Chapter 5 (Die Radfahrtruppe: 1914-1945)

1. U. S. War Department, *Handbook on German Military Forces*, p. 2.

2. Yergin, pp. 179-182, 208, 382, and Chapter 9.

3. *The Times*, Aug. 6, 1919, p. 4; Clark, ed., *The London Cyclist Battalion*, pp. iii-iv.

4. See, for example, the British War Office's *Handbook of the Military-Pedal-Bicycle*, 1933, and *Instructions for the Use of Permanent Staff Instructors of the Supplementary Reserve, Territorial Army and O.T.C., in Regard to the Care, Inspection and Repair, etc., of Small Arms, Machine Guns and Bicycles*, 1933.

5. Ministerio della Guerra, Istruzione sul Materiale Ciclistico, 1927; *Istruzione sull'affardellamento dei Bersaglieri Ciclisti e Motociclisti e sul Trasporto della Armi e delle Munizione sulla Bicicletta e sui Motomezzi*, 1939; R. L. K. Allen, "Cyclist Units," *Journal of the Royal United Service Institution*, vol. 80, 1935, pp. 109-115; Fairfax Downey, "It Wasn't Always Boxcars," *The American Legion Weekly*, Sept. 18, 1925, p. 10; Walter Duranty, "Russian Athletes Having a New Day," *New York Times*, Feb. 2, 1937, p. 37.

6. Allen, "Cyclist Units," pp. 109-115; Major I. R. C. G. Bruce, "Cyclist

Units," *Journal of the Royal United Service Institution*, vol. 80, 1935, pp. 641-642.

7. *Cycling*, Oct. 26, 1916, pp. 336-337; Allen, "Cyclist Units," pp. 110-111; C. D. Eddleman, "Self-Propelled Doughboys," *Infantry Journal*, vol. 45, p. 234; Barbara Tuchman, *The Guns of August*, pp. 49, 54, 171; *The Illustrated War News*, Oct. 21, 1914, p. 23.

8. Much of this section was drawn from information excerpted in R. S. Kohn's study, *Bicycle Troops*, prepared for the Pentagon in 1965 (see the chapter on Vietnam for details). Portions of *Die Radfahrtruppe* were translated in Appendix B, pp. B1-B16, and Appendix D, pp. D26-D38.

9. Klaus Richter, *Weapons and Equipment of the German Cavalry, 1935-1945*, pp. 3, 16-17, 31, 34, 37, 40; Oberleutnant A. D. Furstenberg, *Radfahrfibel Zusammengestellt und Bearbeitet*; Reichskriegministerium, *Das Truppenfahrrad*; Robert O'Neill, personal communication, Canberra, ACT, Australia, 1978.

10. Robert T. Elson, *Prelude to War*, p. 184.

11. Yergin, pp. 328-333, 336-343, 382.

12. Martin Walker, "An Ersatz Army," a review of Albert Seaton's The German Army 1933-1945, in *History Today*, June 1982, pp. 59-60; Matthew Cooper, *The German Army 1933-1945: Its Political and Military Failure*, 1978. Cooper's work, an extensive and detailed analysis of the structure of the German Army and its weaknesses, fails to mention bicycles, bicycle troops, or their use, either during the course of their early successes, or during the later re-structuring of forces and subsequent heavy reliance upon the two wheeled devices.

13. Nicholas Bethell, *Russia Besieged*, pp. 12-13, 64, 173, 189; Bryan Perrett, *A History of Blitzkrieg*, pp. 124-125. Except where otherwise noted, the remainder of this section is based upon information drawn from the U.S. War Department's *Handbook on German Military Forces*. The *Handbook*, containing a vast amount of detailed information, was published in early 1945, while the war was still in progress, and reprinted in 1990 by Louisiana State University Press. As Stephen Ambrose noted in his introduction to the reprint, since it was compiled and written by Americans, who first entered combat against the Germans in 1942 in North Africa, it has some weaknesses with respect to German activities during the early blitzkriegs, before the Americans were involved, and with respect to the Russian front, about which the Russians were reluctant to reveal much. That includes data regarding early German bicycle troop use.

14. U.S. War Department, *Handbook on German Military Forces*, p. 306.

15. *The New York Times*, June 6, 1940, Section VII, p. 2; Leonard Moseley, *The Battle of Britain*, p. 18; *The Army Quarterly*, "Editorial," July 1940, pp. 199-201.

16. "Finns Gather Bicycles to Replace Army's Skis," *New York Times*, Feb. 19, 1940, p. 9.

17. Nicholas Bethell, *Russia Besieged*, p. 8; Comments made to the author by Harrison Salisbury during a personal interview, April 1983; Bryan Perrett, *A History of Blitzkrieg*, 1983, p. 130.

18. Captain Joseph Thorne, "Army Cycles," *Journal of the Royal United Service*

Institution, vol. 80, 1935, pp. 642-643.

19. Richard Collier, *D-Day, 6th June, 1944, The Normandy Landings*, p. 150.

20. As a possible reflection of the general American view towards bicycles, Stephen Ambrose noted that when one 101st Division commander tried to take a bicycle with him, his soldiers had it thrown overboard during the Channel crossing, *June 6, 1944, D-Day, The Climactic Battle of World War II*, pp. 510-511. Information on U. S. airborne cycle trials was cited in a letter from the Military History Research Collection to a private citizen making inquiries about the unit, April 27, 1973, made available to the author. The letter specifically cited John. T. Ellis's work *The Airborne Command and Center: Army Ground Forces Study No. 25*; records of the Organizational History Branch, Office, Chief of Military History, Department of the Army; and the parachute/airborne infantry battalion tables #7-35, Aug. 1, 1944, and #7-35T, Dec. 16, 1944. Bicycle equipment experiments went on elsewhere. In Australia, in 1942, for example, tests were undertaken of carrying 3-inch mortars on bicycles. The conclusion was not news: it was much easier for the soldiers to push six mortar shells, weighing 60 pounds, on bikes than to carry them, by using a commercially made carrier on the rear of the machine. One interesting comment was that much larger saddles were needed than the current narrow, hard civilian models. In particular, wider saddles made it possible to tie a wide variety of equipment underneath the saddle and along the crossbar for more effective transport. That is the kind of useful trivia that usually only turns up in the field, once it is too late to make changes.

21. Photographs were reviewed by the author at the Imperial War Museum and National Army Museum, London.

22. Photographs of allied cycle troops are included in many accounts of the Normandy invasion, ranging from popular writing to specialized military history works. One photo in particular, of cycle troops wading ashore holding onto lines strung between the landing craft and beach, is used in numerous books. Comments on the bicycle's use, however, are restricted to accompanying captions, and are relatively uninformative. Photographs referred to in this section—mostly from the Imperial War Museum—include those reviewed personally by this author, and others reproduced in such works as Stephen Ambrose, *June 6, 1944, D-Day, The Climactic Battle of World War II*, ff. p. 320; Gerald Astor, *June 6, 1944: The Voices of D-Day*, ff. p. 116; Richard Collier, *D-Day, 6th June, 1944, The Normandy Landings*, pp. 113, 129, 189; James D. Ladd, *Inside the Commandos: A Pictorial History from World War Two to the Present*, p. 43; and Cornelius Ryan, *The Longest Day*, p. 221.

23. Ambrose, *June 6, 1944, D-Day, The Climactic Battle of World War II*, pp. 561-562, 571.

24. *The New York Times*, June 7, 1944, p. 5. The German use of the bicycle at even the highest levels of command is evidenced by an interesting incident. In the face of the rapid allied onslaught, a meeting was scheduled among high-ranking German commanders which some suggest was effectively an attempt

to arrange a surrender. On the way to the meeting a small convoy, including the German General Kluge and his son, were attacked by allied aircraft and most of them killed or wounded. The meeting was so crucial that while Kluge lay in a ditch, his assistant grabbed a bicycle to ride on to the scheduled rendezvous. The other commanders, however, had already left the meeting site by the time the assistant arrived.

25. Basil Liddell Hart, ed., *History of the Second World War*, pp. 565-567; Yergin, pp. 384-388; David Irving, *The War Between the Generals*, p. 226.

26. Martin Blemenson, *Liberation*, 1978; "Daisies from the Killing Ground," *Time*, May 28, 1984, p. 21; Herbert Walther, *The Waffen SS: A Pictorial Documentation*, p. 176.

Chapter 6 (When Tojo Came A-Wheeling)

1. Peter Charlton, *War Against Japan 1941-1942*, p. 50.

2. Vincent Freeth, *The Bruce Small Story*, p. 14.

3. "The Sport and Trade in Japan," *The Australian Cyclist*, Jan. 16, 1896, p. 17.

4. Teijiro Uyeda and Hiroshi Koyasu, *Small-scale Industries of Japan: The Bicycle Industry*; the *Japan Statistical Yearbook* (various years); and the *Japanese History of the Census of Manufactures* (various). Translation and interpretation of the latter two were provided by Ahiromitsu Mori.

5. "Britons to Fight Japanese Imports," *New York Times*, Feb. 28, 1934, p. 41; *The Times* (London), Feb. 24, 1934, p. 14.

6. "Cited for Bicycle Sales," *New York Times*, June 20, 1928, p. 8.

7. Personal interview with Edward Riley, former cycle shop owner, Collie, Western Australia, Dec. 11, 1976.

8. Major-General S. Woodburn Kirby, *The Loss of Singapore*, p. 211.

9. Churchill, *My African Journey*, pp. 137-138.

10. Churchill, *The Hinge of Fate*, p. 37.

11. Colonel Masanobu Tsuji, *Singapore: The Japanese Version*, is a basic source of information on the Malayan invasion. Chapter 35, "The Bicyclists," discusses the machine's use in some detail, but is lacking in certain respects; all subsequent references attributed to Tsuji are from that chapter. War History Office, *Mare Shinko Sakusen*, p. 193; and Yoshiko Saito, *Moshin Mare Shingaporu*, pp. 75-79, translations by Ahiromitsu Mori.

12. F. Spencer Chapman, *The Jungle is Neutral*, p. 27; Lieutenant P. W. Thompson, "The Japanese Army, Part 2, The Jungle War in Malaya," 1983; "By an R.A.A.F. Officer," *Great Was the Fall: The Story of a Malayan Tragedy*, p. 98.

13. Chapman, *The Jungle is Neutral*, pp. 27-28.

14. Stanley Falk, *Seventy Days to Singapore: The Malayan Campaign, 1941-1942*, p. 137, a very detailed history by the Japanese-speaking Chief Historian of the United States Air Force; *The Bulletin* (Australia), Jan. 28, 1942, p. 23.

15. Gavin Long, *The Six Years War: A Concise History of Australia in the 1939-1945 War*, p. 140; Lionel Wigmore, *The Japanese Thrust*, pp. 212-216.

16. Long, *The Six Years War: A Concise History of Australia in the 1939-1945 War*, p. 139.

17. Kirby, *The Loss of Singapore*, p. 335; Raymond Callahan, *The Worst Disaster: The Fall of Singapore*, pp. 245-246; Chapman, *The Jungle is Neutral*, p. 124.

18. General Gordon Bennett, in the "Introduction" to Tsuji's book, *Singapore: The Japanese Version*, p. vii; John Keegan, *The Second World War*, p. 259.

19. Keegan, pp. 257-258.

20. Roy Dilley, *Japanese Army Uniforms and Equipment: 1939-1945*, pp. 5, 7; *The Bulletin* (Australia), Jan. 28, 1942, p. 23.

21. Tim Bowden, *Changi Photographer: George Aspinall's Record of Captivity*, p. 42.

22. Observations on the personality and behavior of Colonel Tsuji are from John Toland, *The Rising Sun: The Decline and Fall of the Japanese Empire, 1936-1945;* a personal interview with John Toland on April 5, 1983; Arthur Zich, *The Rising Sun*, p. 123; and Sterling Seagrave, *Lords of the Rim*, chapters 10 and 11.

23. Many publications contain minor written and photographic evidence on the Japanese use of the bicycle, including, for example, Robert Hoare's *World War Two: An Illustrated History*, Arthur Zich's *The Rising Sun*, and Dilley's *Japanese Army Uniforms and Equipment: 1939-1945*.

24. Maurice Collis, *Last and First in Burma*, p. 76; *The Bulletin* (Australia), Feb. 11, 1942, p. 22.

25. Lida Mayo, *Bloody Buna*, p. 21.

26. James D. Ladd, *Inside the Commandos: A Pictorial History from World War Two to the Present*, p. 52.

27. Chapman's book, *The Jungle is Neutral*, is a military history classic (the hardback reprinted fifteen times, followed by continuing paperback reprints) and is basic reading for anyone interested in the wartime use of the bicycle. The incidents cited here are only a small sample of what he relates.

Chapter 7 (The Home Front: World War II)

1. Charles Whiting, *The Home Front: Germany*, pp. 29, 62; *The New York Times*, Oct. 19, 1939, p. 10.

2. *The New York Times*, Aug. 14, 1940, p. 4; Oct. 24, 1940, p. 4; Feb. 1, 1941, p. 5; and March 29, 1943, p. 8.

3. Personal correspondence with Hans Brunswig, Hamburg, Germany, Aug. 12, 1993.

4. Kathleen Cannell, "Parisians Returf Air Raid Trenches," *The New York Times*, July 14, 1940, p. 18; "Bicycle Taxis Face Curb," *The New York Times*, July 3, 1943, p. 5; and *The New York Times*, Dec. 3, 1943, p. 6.

5. Russell Miller, *The Resistance*, p. 46; *The New York Times*, Aug. 29, 1941, p. 6.

6. *The New York Times*, May 9, 1942, p. 5.

7. M. R. D. Foot, *S O E in France: An Account of the Work of the British Special*

Operations Executive in France, 1940-1944, p. 297; Anthony Cave Brown, *Bodyguard of Lies*, pp. 344-346.

8. Blumenson, *Liberation*, presents an excellent picture of the state of the city when liberated. The photo-essay on "The Parisians Master War," pp. 116-127, refers to velo-taxis and man-powered cinema lighting and ventilation systems, among other cycle technology adoptions. See also *The New York Times*, Sept. 3, 1944, p. 8; and Sept. 10, 1944, p. 36.

9. Yergin, *The Prize: The Epic Quest for Oil, Money, and Power*, pp. 370-371; *The Times*, Nov. 2, 1939, cited in Marion Yass, *The Home Front: Britain, 1939-45*.

10. "Bicycle Output Drops," *The New York Times*, Jan. 20, 1940, p. 27; "Mechanical Devices at Premium in Britain," *The New York Times*, Sept. 13, 1942, section IX, p. 3.

11. R. M. Younger, "Bicycles Have a New Importance and Popularity in War-Time," *The Adelaide Advertiser*, Feb. 7, 1942, p. 11; Seamus McGonagle, *The Bicycle in Life, Love, War and Literature*, p. 105.

12. "Britain's Italian Captives Ride Bicycles to Work," *The New York Times*, April 9, 1943, p. 2.

13. H. H. England, *War-Time Cycle Lamp & Lighting Regulations. How to make "Cycling's" lamp mask*, English Universities Press, 1943; *The Times*, Feb. 19, 1943, p. 8, and Dec. 23, 1944, p. 4.

14. Forbes Miller, *Australia Since the Camera: The Second World War*, Colin Forster, *Industrial Development in Australia*, p. 30; S. J. Butlin, *War Economy 1939-1942*, pp. 280-292.

15. A. J. Sweeting, "World War II," *The Australian Encyclopaedia*, p. 135; Michael McKernan. *All In!: Australia During the Second World War*, pp. 87-89.

16. Younger, "Bicycles Have a New Importance," p. 11; "The N.R.M.A. Carries On, Despite War's Effects," *The Open Road*, Oct. 31, 1940.

17. The discussion of Australian cycle manufacturing during the war is based upon information drawn from Vincent Freeth, *The Bruce Small Story*, a commissioned work about the company. Further details about Bruce Small himself are provided in Sir Hubert Opperman, *Pedals, Politics, and People*. Background on General Douglas MacArthur in Australia is from William Manchester, *American Caesar: General Douglas MacArthur, 1880-1964*.

18. Manchester, *The Glory and the Dream: A Narrative History of America, 1932-1972*, pp. 289-294.

19. Yergin, *The Prize*, pp. 371-381.

20. Articles in *The New York Times*, listed below, were the sources of information for the remainder of this section, unless otherwise noted: "Order Trebles Adult Bicycles," March 13, 1942, p. 38; "More Bicycles to Keep Nation on Wheels," March 22, 1942, Section IX, p. 1; Charles E. Egan, "Sale of Bicycles to Adults Curbed by Order of WPB," April 3, 1942, p. 1; "Few Bicycles Left for Rationing Here," April 4, 1942, p. 15; "Topics of the Times," April 6, 1942, p. 14; "Bicycle Order Extended," April 8, 1942, p. 21; "Bicycles," June 4, 1942, p. 29; "'War Model' Bicycle Gets Price

Ceilings," June 6, 1942, p. 11; "Bicycles," June 17, 1942, p. 35; "Bicycles for Adults Ready for Rationing," July 3, 1942, p. 11; "Eligibility of Rules for Bicycles Issued," July 6, 1942, p. 12; "Jersey Defers Cycle Rationing," July 8, 1942, p. 45; "Rationing of Bicycles Begun in New York; Defense and War Workers Can Get Them," July 10, 1942, p. 19; "Adults Buy Up Bicycles," July 25, 1942, p. 26; "Bicycles," July 29, 1942, p. 29; "Cuts Bicycle Buyers to Essential Users," Aug. 11,1942, p. 29; "Bicycles," Aug. 29, 1942, p. 23; "Clarifies Bicycle Ceiling," Dec. 5, 1942, p. 24; "'Waac'-Cycles," Feb. 20, 1943, p. 16; "The End of Rationing," Oct. 24, 1944, Section V, p. 2; and "New Steps Taken for Civilian Needs," May 23, 1945, p. 27.

21. Yergin, *The Prize*, pp. 207-208, 409.

Chapter 8 (The Bicycle in Vietnam)

1. Michael Maclear, *Vietnam: The Ten Thousand Day War*, p. ix.

2. Hugh Lunn, *Vietnam: A Reporter's War*, p. 181; Bernard Fall, *Street Without Joy*, p. 15.

3. Gabriel Kolko, *Anatomy of a War: Vietnam, the United States, and the Modern Historical Experience*, p. 5; Personal interview with Greg Lockhart, Aug. 23, 1993; Bernard Fall, *The Viet—Minh Regime*, p. 155; Maclear, *Vietnam: The Ten Thousand Day War*, p. 20. Maclear's book is derived from a television series in which a large number of people were interviewed. As a consequence, it is a valuable collection of comments on a number of matters, both during the French and American periods of involvement. Jean Lacouture, "Preface," General Vo Nguyen Giap, *Banner of People's War, the Party's Military Line*, pp. vii-viii, x; Douglas Pike, *PAVN: People's Army of Vietnam*, p. 27; Georges Boudarel, "Introduction," in General Vo Nguyen Giap, *Banner of People's War, the Party's Military Line*, p. xiii; General Vo Nguyen Giap, *Dien Bien Phu*, pp. 29-30, italics in original; Douglas Pike, *PAVN: People's Army of Vietnam*, p. 226.

4. Marshall Andrews, "Foreword," in Bernard Fall's *Street Without Joy*, p. 9; Fall, *Street Without Joy*, p. 15; Maclear, *Vietnam*, pp. 27-28.

5. Fall, *Hell in a Very Small Place*, p. ix; Maclear, *Vietnam*, pp. 28-29; *Voice of North Vietnam*, Sept. 22, 1953, cited in Fall, *The Viet-Minh Regime*, p. 84; Andrews, "Foreword," in Fall's *Street Without Joy*, p. 11-12; Claude Guiges, "Logistique Vietnam," *Indochine-rudest Asiatique*, March 1953, cited in Fall, *The Viet-Minh Regime*, p. 85.

6. Fall, *The Viet-Minh Regime*, p. 91.

7. Fall, *Hell in a Very Small Place*, p. 451; Maclear, *Vietnam*, p. 38.

8. Vien Su Hoc, *May Van De*, pp. 97, 108-115, 174-176, 196-198, cited in Greg Lockhart, *Nation in Arms: The Origins of the People's Army of Vietnam*, pp. 259-260; Fall, *Hell in a Very Small Place*, pp. 128-129, 131; LSQD, p. 558, on keeping the roads open, cited in Lockhart, *Nation in Arms*, p. 260.

9. Nguyen An's essay on the transport system in Vien Su Hoc, *May Van De*, p. 174, cited in Lockhart, *Nation in Arms*, p. 260; Fall, *Hell in a Very Small Place*, p. 452. The Vietnamese figures are from, for example, Vien Su Hoc, *May Van*

De, p. 97, cited in Lockhart, *Nation in Arms,* p. 259. Lockhart, in a personal interview on June 23, 1993, said that he found it very difficult to work out the statistics of how much rice and other supplies were needed to support a guerrilla unit in the field. When he got down to details, he found the available information inconsistent. His final assessments were impressionistic as, I suspect, are most others that are cited. Edgar O'Ballance, *The Indo-China War (1945-1954): A Study in Guerrilla Warfare,* pp. 112-113, whose figures are reproduced here, said only that "it was estimated," but not by whom.

10. Maclear, *Vietnam,* p. 32.

11. Several of the authors cited in this chapter (as well as many others not cited) have stated at one or more points in their writing the "typical" weight carried by the Vietnamese cycle porters. I have not listed the sources and page numbers here, as none of them actually say where the figures come from. Most of them, I suspect, are impressionistic or repetitions of what others have said.

12. Fall, p. 452; Lockhart, *Nation in Arms,* p. 260, citing Tran Do, *Recits sur Dien Bien Phu, p.* 13.

13. Fall, p. 452.

14. Nguyen An's essay on the transport system in Vien Su Hoc, *May Van De,* p. 174, cited in Lockhart, p. 260; Bernard Fall, p. 451-452.

15. Bernard Fall, p. 127.

16. Maclear, chapter 4.

17. Thomas C. Thayer, "Patterns of the French and American Experience in Vietnam," pp. 37-38. Several Australian officers who served in Vietnam conveyed to the author that while they personally liked the Americans, they found in working with them that they had a "propensity not to learn from anybody," as one put it.

18. Fall, *Street Without Joy,* pp. 355, 381; V. J. Croizat, *Lessons of the War in Indochina, p.* iii.

19. Lunn, *Vietnam: A Reporter's War,* p. 10; Fall, *Hell in a Very Small Place,* p. 458-459.

20. Maclear, p. 156; Lunn, p. 37.

21. Michael Herr, *Dispatches,* p. 133.

22. Boudarel, "Introduction," in General Vo Nguyen Giap, *Banner of People's War, the Party's Military Line,* p. xxiii; Truong Nhu Tang, *Journal of a Vietcong,* p. 166.

23. Giap, *Banner of People's War, the Party's Military Line,* p. 67; Boudarel, "Introduction," in General Vo Nguyen Giap, *Banner of People's War, the Party's Military Line,* pp. xvi, xviii; Gabriel Kolko, *Anatomy of a War: Vietnam, the United States, and the Modern Historical Experience,* p. 183.

24. Kolko, *Anatomy of a War: Vietnam, the United States, and the Modern Historical Experience,* p. 185; Fall, *The Viet-Minh Regime,* p. 155; Douglas Pike, *PAVN: People's Army of Vietnam,* pp. 58-59.

25. Lunn, p. 89.

26. Burchett is a highly controversial figure in Australia, considered by some

to be a "grubby traitor," as one reader of this manuscript put it, because of his actions in the Korean War. Nonetheless, as another reader noted, he was one of the very few westerners to write from the communist Vietnamese perspective, and their trust of him led Henry Kissinger to use him as a liaison during early, unofficial approaches to the North. His notes on the bicycle's use were not matched by any other western writer of the time.

27. Wilfred G. Burchett, *Vietnam: Inside Story of the Guerrilla War*, pp. 25-26. His comments on the use of the bicycle are found on pages 16-17, 25-28, 31, 48-49, 56, and 61-62. Those wanting to follow up will find the book of considerable interest.

28. Truong Nhu Tang, *Journal of a Vietcong*. A founding member of the National Liberation Front of South Vietnam, and Minister of Justice in the Provisional Revolutionary Government, Tang became disillusioned after the communist takeover and fled Vietnam. Cycling references are from pages 128, 158-162, 168, and 181.

29. Harrison Salisbury, "North Vietnam Runs on Bicycles," *The New York Times*, Jan. 7, 1967, pp. 1, 3; *Behind the Lines-Hanoi: December 23, 1966-January 7, 1967*.

30. See, for example, Seamus McGonagle, *The Bicycle in Life, Love, War and Literature*, p. 10, citing English newspaper coverage.

31. R. S. Kohn, *Bicycle Troops*, p. 1; "Report of Evaluation Tote Gote Test in Thailand, Summary Report, Vol. I," Joint Thai-U. S. Military Research and Development Center, Bangkok, Thailand, 64-014, Dec. 1964.

32. Lieutenant Colonel P. W. Thompson, "The Japanese Army, Part 2, Jungle War in Malaya," *Infantry Journal*, May 1943, pp. 10^1-5; R. S. Kohn, *Bicycle Troops*, pp. 3, 19, 22, 29, 48, citing Jules Roy, *The Battle of Dienbienphu*; Department of the Army. "Counterguerrilla Operations," *Department of the Army Field Manual. FM 31-16*, Feb. 1963; Department of the Army, "U. S. Army Counterinsurgency Forces," *Department of the Army Field Manual FM 31-22*, Nov. 1963; and RVNAF, *War Material Used by Vietcong in South Vietnam, Handbook II*, J2, High Command, undated.

33. Personal interviews with R. S. Kohn, April 5, 1983, and Harrison Salisbury, April 6, 1983. The attempt to better understand mobility issues in Vietnam may be further reflected in an eight-page memo, dated Aug. 11, 1965, titled *Employment of Bicycles*, and prepared by the U. S. Army Concept Team in Vietnam. Unfortunately, I have been unable to locate a copy.

34. Stanley Karnow, *Vietnam: A History*, p. 331; Truong Nhu Tang, *Journal of a Vietcong*, pp. 240-241.

35. Kolko, *Anatomy of a War: Vietnam, the United States, and the Modern Historical Experience*, pp. 147-148; Karnow, *Vietnam: A History*, p. 332; Maclear, *Vietnam: The Ten Thousand Day War*, p. 174.

36. James S. Olson, *Dictionary of the Vietnam War*, pp. 204-205; Edward J. Marolda and G. Wesley Pryce III, *A Short History of the United States Navy and the Southeast Asian Conflict, 1956-1975*, cited in Olson, *Dictionary of the Vietnam*

War, p. 341.

37. Olson, *Dictionary of the Vietnam War,* p. 218; William E. LeGro, *Vietnam from Ceasefire to Capitulation,* cited in James S. Olson, *Dictionary of the Vietnam War,* p. 218; Pike, *PAVN: People's Army of Vietnam,* p. 47.

38. Olson, *Dictionary of the Vietnam War,* pp. 68, 204-205; Robert W. Chandler, *War of Ideas: The U. S. Propaganda Campaign in Vietnam,* cited in Olson, *Dictionary of the Vietnam War,* pp. 380-381; Pike, *PAVN: People's Army of Vietnam,* p. 47; Kolko, *Anatomy of a War: Vietnam, the United States, and the Modern Historical Experience,* p. 190—191.

39. Karnow, *Vietnam: A History,* p. 456; Truong Nhu Tang, *Journal of a Vietcong,* p. 242.

40. Quoted in Maclear, *Vietnam: The Ten Thousand Day War,* p. 172; Theh Chongkhadikij, "The Ho Chi Minh Highway," *Bangkok Post, Sunday Magazine,* Aug. 25, 1974, pp. 14-15; Truong Nhu Tang, *Journal of a Vietcong,* pp. 241-242.

41. From Salisbury, "North Vietnam Runs on Bicycles," *The New York Times,* Jan. 7, 1967, pp. 1, 3; and *Behind the Lines-Hanoi: December 23, 1966-January 7, 1967.*

42. Michael Richardson, "Wheel to Wheel in Hanoi," *Sydney Morning Herald,* January 5, 1974, p. 8; David Marr, interview with author, Canberra, A.C.T., Australia, Jan. 12, 1979.

43. Truong Nhu Tang, *Journal of a Vietcong,* p. 243.

44. *The New York Times,* July 15, 1967, p. 3; *The New York Times,* Jan. 17, 1968.

Chapter 9 (Retrospect and Prospect)

1. BBC, Europe Region, Friday, 11 May, 2001, 14:15 GMT.

2. Major-General W. H. Mackinnon, *The Journal of the C.I.V. in South Africa; The City Press C.I.V. War Souvenir No. 3: Pictures of the C.I.V. at the Cape, Portaits and Groups,* London: W. H. and L. Collingridge, 1900; J. Barclay Lloyd, *One Thousand Miles with the C.I.V.*

3. Tadao Nakada, *Imperial Japanese Army and Navy Uniforms and Equipment.*

4. Peter Charlton, *War Against Japan 1941-1942, Australians at War,* Time-Life Books. There are numerous other works that ignore the bicycle. These just happen to be excellent examples.

5. *New York Times,* Jan. 30, 1976, p. 1.

6. George Russell, "We Can Move Anywhere," *Time,* March 15, 1982, pp. 6-7.

7. F. P. U. Croker, "The Man-Powered Military Vehicle," *The Army Quarterly and Defence Journal,* vol. 101, no. 4, July 1971, pp. 475-478; Major Stephen Tate, "Human Powered Vehicles in Support of Light Infantry Operations", Master of Military Art and Science thesis, Faculty of the U.S. Army Command and General Staff College, Fort Leavenworth, Kansas, 1989.

8. J. F. C. Fuller, *Memoirs of an Unconventional Soldier,* pp. 462-463.

9. Lieutenant Colonel Ron Reid Daley, *Selous Scouts: Top Secret War,* pp. 170-172.

10. Montague Bikes, and a personal interview with David Montague.

11. Noah Shachtman, "Election Race? First, Check Out This Bike," *New York Times*, September 11, 2003.

Sources of Illustrations

The illustrations for this book were collected from a variety of libraries, museums, archival records, military history institutes, books, newspapers, magazines, cycle journals, and photographs loaned to me by interested individuals. I would like to thank everyone who gave me permission to reproduce illustrations from their collections.

Several of the photographs raise intriguing questions. For example, nothing more could be determined about the circumstances surrounding the pictures of the soldiers carrying folding bikes on the Russian front in World War I, or the Russian cycle troops leading dogs in World War II. I would appreciate hearing from anyone with further information.

Page

- Title Page.   Trustees of the Imperial War Museum, London.
- 2. Courtesy Smithsonian Institution Traveling Exhibition Service.
- 3. Courtesy Smithsonian Institution Traveling Exhibition Service.
- 4. Courtesy Smithsonian Institution Traveling Exhibition Service.
- 5. Courtesy Smithsonian Institution Traveling Exhibition Service.
- 6. *The Engineer*, February, 1888.
- 7. Courtesy Smithsonian Institution Traveling Exhibition Service.
- 8. *The Austral Wheel*, January, 1897.
- 9. *The Lone Hand*, March 1, 1911.
- 10. *The Austral Wheel*, February, 1897.
- 12. Dunlop Rubber (Australia) Ltd., Noel Butlin Archives, Australian National University, Canberra.
- 13. Jack Costello, Rivervale, West Australia.
- 14. *The Strand*, 1891.
- 15. Courtesy Smithsonian Institution Traveling Exhibition Service.
- 17. *The Strand*, 1891.
- 22/23. U.S. Army Military History Institute.
- 26/27. K. Ross Toole Archive, The University of Montana, Missoula, photo #73-31.
- 29. U.S. Army Military History Institute.
- 30. U.S. Army Military History Institute.

32. *Scientific American*, December 28, 1901.
33. (Top) *Scientific American*, December 28, 1901.
    (Bottom) Richard Fisher.
34. (Top) A. S. Clark (editor), *The London Cyclist Battalion*.
    (Bottom) Courtesy Smithsonian Institution Traveling Exhibition Service.
35. A. S. Clark (editor), *The London Cyclist Battalion*.
37. *Scientific American Supplement, no. 1220*, May 20, 1899.
39. *The Sketch*, September 30, 1896.
41. H. W. Wilson, *After Pretoria*, vol. 1.
44. Origin unknown.
46. A. S. Clark (editor), *The London Cyclist Battalion*.
48. Courtesy Smithsonian Institution Traveling Exhibition Service.
49. Colin Crisswell, *Far East History, 1870-1952*.
50. H. O. Duncan, *The World on Wheels*, vol. 1.
52. H. W. Wilson, *With the Flag to Pretoria*, vol. 1.
53. *The Austral Wheel*, August, 1896.
55. H. W. Wilson, *With the Flag to Pretoria*, vol. 2.
57. (Top) H. W. Wilson, *With the Flag to Pretoria*, vol. 2.
    (Bottom) H. W. Wilson, *With the Flag to Pretoria*, vol. 2.
58. H. W. Wilson, *With the Flag to Pretoria*, vol. 2.
59. *Black and White Budget*, March 31, 1900.
60. H. W. Wilson, *With the Flag to Pretoria*, vol. 2.
62/63. *Cape Times Weekly Edition*, July 30, 1902, Independent News Cape.
64. War Museum, Bloemfontein.
65. H. W. Wilson, *With the Flag to Pretoria*, vol. 1.
66. *Cape Times Weekly Edition*, November 7, 1900, Independent News Cape.
68. (Top) J. Barclay Lloyd, *One Thousand Miles with the C.I.V.*
    (Bottom) H. W. Wilson, *With the Flag to Pretoria*, vol. 2.
69. *Cape Times Weekly Edition*, September 19, 1900, Independent News Cape.
71. H. W. Wilson, *With the Flag to Pretoria*, vol. 2.
72. H. W. Wilson, *With the Flag to Pretoria*, vol. 1.
73. *Cape Times Weekly Edition*, February 6, 1901, Independent News Cape.
74. *Cape Times Weekly Edition*, February 6, 1901, Independent News Cape.
76. *Cape Times Weekly Edition*, October 25, 1899, Independent News Cape.
78. H. W. Wilson, *With the Flag to Pretoria*, vol. 2.
80. Dunlop Rubber (Australia) Ltd., Noel Butlin Archives, Australian National University, Canberra.
81. Dunlop Rubber (Australia) Ltd., Noel Butlin Archives, Australian National University, Canberra.
82/83. Courtesy of the Director, National Army Museum, London.
87. Western Australian Museum, Perth.
92. Origin unknown.
93. Sectie Militaire Geschiedenis Landmachtstaf (Holland).
94. Origin unknown.

96/97. *Cycling*, August 10, 1916.
98/99. The Trustees of the Imperial War Museum, London.
101. Mrs. Cecily Adams, Castlecrag, New South Wales.
102. Mrs. Cecily Adams, Castlecrag, New South Wales.
104. Australian War Memorial, negative number H02943.
105. The Trustees of the Imperial War Museum, London.
106. Dunlop Rubber (Australia) Ltd., Noel Butlin Archives, Australian National University, Canberra.
108. (Top) Origin unknown.
(Bottom) Wikipedia, Lombardi Historical Collection.
109. *Cycling*, July 26, 1917
110. *Cycling*, December 13, 1917.
112. Sectie Militaire Geschiedenis Landmachtstaf (Holland).
113. Sectie Militaire Geschiedenis Landmachtstaf (Holland).
114. Sectie Militaire Geschiedenis Landmachtstaf (Holland).
116. Mrs. Cecily Adams, Castlecrag, New South Wales.
117. Mrs. Cecily Adams, Castlecrag, New South Wales.
118. *Cycling*, April 19, 1917.
119. Origin unknown.
120. *Cycling*, March 22, 1917.
121. The Trustees of the Imperial War Museum, London.
123 The Trustees of the Imperial War Museum, London.
124. The Trustees of the Imperial War Museum, London.
125. Origin unknown.
126. The Trustees of the Imperial War Museum, London.
127. Origin unknown.
128. Bundesarchiv.
129. The Trustees of the Imperial War Museum, London.
130/131. The Trustees of the Imperial War Museum, London.
132. The Trustees of the Imperial War Museum, London.
134. Courtesy of the Director, National Army Museum, London.
135. The Trustees of the Imperial War Museum, London.
136. The Trustees of the Imperial War Museum, London.
138. The Trustees of the Imperial War Museum, London.
139. The Trustees of the Imperial War Museum, London.
140. The Trustees of the Imperial War Museum, London.
142. Origin unknown.
144. Origin unknown.
145. Australian War Memorial, negative #127909.
146/147. Australian War Memorial, negative #127897.
148. Origin unknown.
151. Origin unknown.
153. Courtesy of *The Sankei Shimbun*.
154. Australian War Memorial, negative number unknown.

Sources of Illustrations

158. Hans Brunswig, Hamburg.
159. The Trustees of the Imperial War Museum, London.
160. The Trustees of the Imperial War Museum, London.
161. The Trustees of the Imperial War Museum, London.
162/163. The Trustees of the Imperial War Museum, London.
167. Malvern Star Bicycles, Australia.
168. (Top) NRMA, *Open Road*, Sydney.
     (Bottom) Malvern Star Bicycles, Australia.
169. Reproduced in numerous Australian newspapers.
170. Malvern Star Bicycles, Australia.
174. The Trustees of the Imperial War Museum, London.
176. Military History Institute, Hanoi.
181. Military History Institute, Hanoi.
182/183. Military History Institute, Hanoi.
184. Military History Institute, Hanoi.
191. Permission of International Publishers Co., Inc., New York.
198. Permission of International Publishers Co., Inc., New York.
199. Military History Institute, Hanoi.
206. (Top) Swiss Armee-Photodienst.
     (Bottom) Swiss Armee-Photodienst.
208. Jim Fitzpatrick.
211. Montague Bikes.
212. Montague Bikes.
213. Montague Bikes.

Acknowledgements

During the late 1970s I carried out a study of the use of the bicycle in rural Australia between 1890 and 1915, going through an immense amount of international cycling literature and archival materials. In the process I ran across various references to the bicycle's military use and wanted to explore the matter further but, living in Australia, thought I would not be able to do much about it. However, it gradually became clear that of the four most significant uses of the bicycle in wartime—the Boer War, World War I, the Japanese invasion of Malaya, and Vietnam—only Anzac forces had been involved in all of them. The excellent libraries and archival sources of the Australian War Memorial, the Commonwealth of Australia, The Australian National University, and the National Library of Australia all lay within a few kilometers of my house. Thus, in several respects, I found myself as close to the center of military cycling history as one could wish to be. Over the next two decades (it was a part-time hobby) I was assisted and encouraged by numerous people from around the world who found the topic fascinating.

    I was also aided by the Australian War Memorial, whose Trustees gave me a small grant towards a trip to England and the United States to work in those countries' libraries, archives and military museums. I wish to acknowledge their generosity.

    Many played a role in getting this book off the ground. Gavan Daws, Canberra and Honolulu, the late Sir Hubert Oppermann, Melbourne, and the late Eric Woolmington, head of the Department of Geography at the Royal Military College, Duntroon, strongly supported the project from the outset. Sylvie Shaw, ABC Radio (Australia), produced a program in 1985 on the bicycle's military use, based upon my work to date. And Margaret Barca, *This Australia,* and the late Ken Henderson, *Hemisphere,* published early articles. David Wilson, Emeritus professor of Engineering at M.I.T. and the author of *Bicycling Science,* read the entire manuscript and offered critical advice. Don McKuen, of Brassey's, Inc., published the first edition in 1998, and my wife, Roey, did the design and layout for both the original and this 2011 revised edition.

Brian Bond, Department of War Studies, Kings College, London, invited me to speak to his students on the role of the bicycle in wartime. Len Deighton, London, gave advice and offered leads. Ian Jones, Melbourne, discussed mutual research interests about Anzacs (especially the Australian Light Horse, from which were recruited a number of Anzac cyclists). Thomas Pakenham, London, helped unearth new leads on the use of the bicycle in the Boer War. Hugh Lunn, Brisbane, read the chapter on Vietnam; as a Reuters reporter at the time of the Tet Offensive, he had some especially valuable comments. Ian McNeill, Canberra, and Greg Lockhart, Senior Research Associate, Faculty of Asian Studies at The Australian National University, also offered thoughts and read the chapter on Vietnam. The late R. S. Kohn, Battelle Memorial Institute, Columbus, Ohio, discussed with me that organization's study of the bicycle's wartime use that was done for the Pentagon. Warren Lennon, Canberra, related his experiences in Vietnam and his views of Australian military attitudes toward the Vietnamese and approaches to the war. The late Harrison Salisbury, of *The New York Times*, talked enthusiastically about the bicycle's use in World War II and Vietnam. The late John Toland offered several insights on the Japanese use of bicycles, and read the chapter on the invasion of Malaya-Singapore. David Montague, of Montague Bikes, provided information, photos and comments for the final chapter of this revised edition.

Generous research assistance was provided by Jane Carmichael, Imperial War Museum, London; Mrs. Marion Harding, National Army Museum, London; the staff of the Melbourne State Library; Sue McKemmish, Australian Archives, Brighton, Victoria; Donald Berkebile, Donald Kloster, Andrea Stevens and Roger White, of the Smithsonian Institution, Washington, D.C; and Mike Saclier, Noel Butlin Archives of Business and Labour, Australian National University, Canberra, and his staff for extensive assistance into all aspects of my cycle history research over the years. Above all, a special thanks to the staff at The Australian War Memorial, Canberra, including then-Deputy Director Dr. Michael McKernan, T. Michael Bogle, Jim Heaton, Michael Piggott, Dr. Peter Stanley and Glynis Vincent. They assisted, supported and encouraged me through it all.

Robert Wallace, Sydney, provided copies of soldiers' personal diaries and records re Australian Boer War cycling activities. The late Cecily Adams, Castlecrag, New South Wales, loaned her father's personal World War I diaries. B. V. Giuliano, N.R.M.A., Sydney, sent pictures of the NRMA WWII road patrols; General Huang Phong, Director, Military History Institute, Hanoi, provided several photographs; Betty Smith, International Publishers, New York, dug out photos of Wilfred Burchett, and Hans Brunswig allowed me to reproduce the photo he took in Hamburg in 1943. Sissi Campbell, Phoenix, Arizona, translated German materials and Ahiromitsu Mori and David Nakamura, Canberra, A.C.T., translated Japanese documents.

To all, a sincere thank you.

Bibliography

All sources with a cited or known author are listed under the author's name. This includes books; newspaper, magazine and journal articles; personal diaries and letters; interviews; and government files and records. Sources without a known author are listed either under official bodies publishing the material (e.g., War Department), or under "Anonymous," by title.

Acting Secretary, Department of Defence. Memo to the Secretary, Department of External Affairs, re "Officers of the Permanent Military Forces of the Commonwealth undergoing Courses of Instruction at the Staff College," Australian Department of Defence file 1862/3/136, May 29, 1911.

Albemarle, the Earl of; and G. Lacy Hillier. *Cycling*, London: Longmans, Green and Co., 1896.

Alderson, Frederick. *Bicycling: A History*, New York: Praeger Publishers, 1972.

Allen, Captain R. L. K. "Cyclist Units," *Journal of the Royal United Service Institution*, Vol. 80, 1935, pp. 109-115.

Altham, Major E. A. (collator). "Notes on the Lines of Communications in the Cape Colony," War Office: Intelligence Division, June, 1899.

Ambrose, Stephen. *June 6, 1944, D-Day, The Climactic Battle of World War II*,

Amery, L. S., ed. *The Times History of the War in South Africa*, London: Sampson Low, Marston and Company Ltd., 1902.

Andrews, Marshall. "Foreword," *Street Without Joy*, London: Pall Mall Press, 1964, pp. 9-14.

Anonymous. "1896 Australian Relay Ride," *Australian Cyclist*, October 12, 1896, pp. 27-40.

—"A Cycling Corps of "Regulars"," *The Cyclist*, June 1, 1898, p. 604.

—"A Folding Bicycle," *Scientific American*, March 13, 1897, p. 165.

—"A Manual of Cyclist Drill," *The Saturday Review*, January 4, 1890, pp. 17-18.

—"A New Military Folding Bicycle," *Scientific American*, October 20, 1900, p. 244.

—"A Preliminary Proposal for Very Lightweight, Off-Road Ground Vehicle Concepts and a Program of Prototype Developments," Cornell Aeronautical Laboratory, Inc. Proposal No. 411, April 24, 1962.

—"A Railway Cycle," *Official Journal of the Cape Town Cycling and Athletics Club*, New Series, No. 12, March, 1900, pp. 7-8.

—"Adults Buy Up Bicycles," *New York Times*, July 25, 1942, p. 26.

—"Army Bicycles," *Scientific American*, August 18, 1894, p. 100.

—"Army Cyclists," *Journal of the Society for Army Historical Research*, Vol. 38, No. 153, March 1960, p. 43.

—"Bicycle Artillery," *Scientific American*, December 28, 1901, p. 21735.

—"Bicycle Order Extended," *New York Times*, April 8, 1942, p. 21.

—"Bicycle Output Drops," *New York Times*, January 20, 1940, p. 27.

—"Bicycle Taxis Face Curb," *New York Times*, July 3, 1943, p. 5

—"Bicycles," *New York Times*, June 4, 1942, p. 29.

—"Bicycles," *New York Times*, June 17, 1942, p. 35.

—"Bicycles," *New York Times*, July 29, 1942, p. 29.

—"Bicycles," *New York Times*, August 29, 1942, p. 23.

—"Bicycles for Adults Ready for Rationing," *New York Times*, July 3, 1942, p. 11

—"Bicycles in State Militia Drill," *Scientific American*, March 23, 1895, p. 182.
—"Bicycles Long Used for Military Service," *The Bicycling World and Motorcycle*, October 6, 1914, p. 36.
—"Bicycling in the Austrian Army," *Scientific American Supplement*, No. 1121, June 26, 1897, pp. 17911-17912.
—"Biking Into Battle," *Army*, June, 1981, pp. 54-55.
—"Birth of the Bike - As Army Transport," *Army*, March, 1979, p. 4.
—"Boer War," *The Australian Encyclopaedia*, Sydney: The Grolier Society, 1983, Vol. 2, pp. 44-46.
—"Britain to Fight Japanese Imports," *New York Times*, February 28, 1934, p. 41.
—"Britain's Italian Captives Ride Bicycles to Work," *New York Times*, April 9, 1943, p. 2.
—"British Spies on Bicycles," *The Standard and Diggers" News*, October 21, 1899, p. 3.
—"Britons to Fight Japanese Imports," *New York Times*, February 28, 1934, p. 41.
—"Cited for Bicycle Sales," *New York Times*, January 20, 1928, p. 8.
—"Clarifies Bicycle Ceiling," *New York Times*, December 5, 1942, p. 24.
—"Crack Cyclist Arrested," *The Cape Times*, January 7, 1896, p. 5.
—"Cuts Bicycle Buyers To "Essential" Users," *New York Times*, August 11, 1942, p. 29.
—"Cycle Infantry," *Hobbies*, June, 1946, p. 21.
—"Cycle Soldiering," *The Australian Cyclist*, January 23, 1896, p. 21.
—"Cycle Telegraphs in War," *Scientific American Supplement*, August 29, 1896, p. 187.
—"Cycling and Home Defence," *The Times*, May 4, 1901, p. 3.
—"Cyclist Infantry," *The Saturday Review*, August 2, 1890, pp. 135-136.
—"Cyclists" Work in Modern Warfare," *Cycling*, May 23, 1918, pp. 392-393
—*Die Radfahrtruppe*, Nach Kriegserfahrungen Bearbeitet von Major Rudolf Theiss im osterr. Radf.-Batl. No. 2, ehemals k.u.k. Radf. Jagerhauptmann und Stabshauptmann Dr. Oskar Regele im osterr. Radf.-Batl. No. 2, Verlag von R. Eisenschmidt, Verlagsbuchhandlung Militarwissenschaft, Berlin, 1925.
—Editorial, *The Army Quarterly*, July, 1940, pp. 199-201.
—"Eligibility of Rules for Bicycles Issued," *New York Times*, July 6, 1942, p. 12.
—"Experimental Bicycle Trip in the United States Army," *Scientific American*, June 5, 1897, p. 354.
—"Few Bicycles Left for Rationing Here," *New York Times*, April 4, 1942, p. 15.
—"Final Report on Test of bicycle, Lightweight, Folding, Model HF-777 (Huffman Manufacturing Company)," ATI 79076, Proof Department, Army Air Forces Proving Ground Command, Eglin Field, Florida, March 24, 1944.
—"Fort Harrison, Mont.," *Army and Navy Journal*, September 5, 1896, p. 5.
—"French Military Folding Bicycle" *Scientific American*, October 26, 1901, p. 264.
—"French Wheelmen," *Army and Navy Journal*, November 14, 1896, p. 182.
—"Finns Gather Bicycles to Replace Army's Skis," *New York Times*, February 19, 1940, p. 9.
—"German Tactics of Combating Guerrillas," *Military Review*, Vol. 24, No. 3, June, 1944, pp. 104-106.
—"Jersey Defers Cycle Rationing," *New York Times*, July 8, 1942, p. 45.
—"Mechanical Devices at Premium in Britain," *New York Times*, September 13, 1942, Section IX, p. 3.
—"Military Bicycling," *Scientific American*, September 19, 1891, p. 185.
—"Military Bicycling," *Scientific American*, November 2, 1895, p. 283.
—"Military Cycling," *The Saturday Review*, June 16, 1888, p. 722
—"Military Cycling," *The Australian Cyclist*, February, 1896, p. 9.

—"Military Intelligence," *The Times*, April 12, 1909, p. 5.

—"More Bicycles to Keep Nation on Wheels," *New York Times*, March 22, 1942, Section IX, p. 1.

—"New Army Bicycles," *Scientific American*, February 8, 1896, p. 91.

—"New Steps Taken for Civilian Needs," *New York Times*, May 23, 1945, p. 27.

—"Note on the Cyclist Brigade [Germany]," memorandum prepared by M.I.2., Positive Branch, Military Intelligence Division, General Staff, November 23, 1918.

—"Order Trebles Adult Bicycles," *New York Times*, March 13, 1942, p. 38.

—"Parisians Returf Air Raid Trenches," *New York Times*, July 19, 1940, p. 18.

—"Parliament and Volunteer Cyclists," *Cycling Times*, January 23, 1896, p. 18.

—"Permits for Cyclists," *The Standard and Diggers" News*, October 14, 1899, p. 5.

—"Rationing of Bicycles Begun in New York; Defense and War Workers Can Get Them," *New York Times*, July 10, 1942, p. 19.

—"Report of Evaluation Tote Gote Test in Thailand, Summary Report, Vol. I," Joint Thai-U.S. Military Research and Development Center, Bangkok, Thailand, 64-014, December, 1964.

—"Sale of Bicycles to Adults Curbed by Order of WPB," *New York Times*, April 3, 1942, p. 1.

—"The Betrayal of Dr. Jim: A Story of the Transvaal Cyclist Scouts," *Cycling Times*, April 23, 1896, p. 8.

—"The Beginnings of War-Machines: Military Bicycles," *The Illustrated War News*, February 21, 1917, pp. 10-11.

—"The Bicycle in the Army," *Scientific American Supplement*, No. 1086, October 24, 1896, p. 17358.

—"The Bicycle in War," *The Austral Wheel*, June 1897, p. 188.

—"The Bicycle Relay Race Across the Continent," *Scientific American*, September 19, 1896, p. 238.

—"The Cycle and the Service," *The Sketch Cycling Supplement, Stanley Show*, September 30, 1896, p. 407-409.

—"The Cycle in Warfare," *The Times*, December 17, 1908, p. 9.

—*The Cyclist Infantry*, August 2, 1890, pp. 135-136.

—"The Easter Cyclist Manoeuvres," *The Saturday Review*, April 7, 1888, pp. 410-411.

—"The End of Rationing," *New York Times*, October 24, 1944, Section V., p. 2.

—"The Folding Military Bicycle," *Scientific American Supplement*, No. 1044, January 4, 1896, pp. 16681-16682.

—"The Folding Military Bicycle" *Scientific American Supplement*, No. 1220, May 20, 1899, p. 19559.

—"The Military Bicycle," *Scientific American*, November 30, 1889, p. 343.

—*The New South Wales Contingent for South Africa*, Sydney: The New South Wales Bookstall Co., 1900.

—"The N.R.M.A. Carries On, Despite War's Effects," *The Open Road*, October 31, 1940.

—"The Sport and Trade in Japan," *The Australian Cyclist*, January 16, 1896, p. 17.

"The War Bicycle," *Scientific American*, March 23, 1897, p. 187.

—"The War Office and Cycles," *The Cyclist*, August 3, 1898, p. 859.

—"The Wheels of Pleasure Spin Again in London," *New York Times*, April 21, 1943, p. 2.

—"Topics of the Times," *New York Times*, April 6, 1942, p. 14.

—"Unfit for Military Service," *The Cyclist*, March 23, 1898, p. 313.

—(Untitled article on military cycling), *The Chautauquan*, February 1897, p. 568.

—(Untitled article on military cycling), *Cycling Times*, January 23, 1896, p. 18.

—(Untitled article on Australian relay ride), *The Australian Cyclist, Special Supplement*, May 21, 1896.

—""WAAC"-Cycles," *New York Times*, February

20, 1943, p. 16.
—"War Material Used by Vietcong in South Vietnam, Handbook II," J2, High Command, RVNAF.
—""War Model" Bicycle Gets Price Ceilings," *New York Times*, June 6, 1942, p. 11
—"War on Wheels," *Black and White Budget*, August 18, 1900, p. 638.
—"With The Cyclist Corps at Easter," *The Cyclist*, April 13, 1898, pp. 404-406
Ansot, H. "A Modern Centaur: A Chapter on Bicycles," *Overland Monthly*, October, 1893, pp. 121-129.
Armstrong, Percy. "Cyclists for the Transvaal," *The Western Mail*, February 17, 1900, pp. 18-19.
Aronson, Sidney, "The Sociology of the Bicycle," *Social Forces*, Vol. 30, 1952, pp. 305-312.
Australian Imperial Force. *Notes on the Belgian Army*, Melbourne: Government Printer, undated (post-1906).
—"Carrying 3 inch Mortar on Bicycle. Result of Test," 975/1/2, May 22, 1942.
Australian War Diaries (Australian War Memorial).
—"1 ANZAC Cyclist Bn.," May 16, 1916, - December 17, 1917.
—"2nd ANZAC Cyclist Battn.," July, 1916 - May, 1917.
—"2nd ANZAC Cyclist Battn.," June, 1917 - December, 1917.
—"2 Div. Cyclist Coy.," March, 1916 - May, 1916.
—"5 Div. Cyclist Coy.," April, 1916 - June, 1916.
—"Aust Corps Cyclist Bn.," January 18, 1918 - April 19, 1918.
Astor, Gerald. *June 6, 1944: The Voices of D-Day*, New York: St. Martin's Press,
Australian War Memorial.
—"1st Anzac Cyclist Battalion. Showing number of cycles with Battalion with men on detailed duties and loaned to other units," NC 235/1, November 4, 1916.
—"Census of Bicycles. 13th A.L.H. Regiment," NC 235/1, January 2, 1917.
—"Census of Cycles. 1st ACB," NC 235/1, January 5, 1917.

—"Establishment of Cyclist Companies. A.I.F.," NC 327/5 [pr 10], March 11, May 23, 1916.
—"Extensive correspondence between the 13th Australian Light Horse Regiment and 1st Anzac Cyclist Battalion as to the ownership and location of bicycles," NC 103/3, April, 1916-October, 1917.
—"Report on Rifle Clips and Carriers on Cycles," NC 235/1, August 9, 15, 1916.
—"Substitution of Bicycles for Motor Cycles in Tractor Drawn Siege Artillery Batteries," NC 103/1, June 28, 1917.
—"Tests carried out of the Carrier Lewes Gun for attachment to Bicycles with results," NC 385/23, May 1, 1916.
—"War Establishments. Mounted Units," NC 327/5 [pr 12], December 6, 1915 / March 1, May 9, July 6, 1916 / March 18, 1917 / February 12, 1918.
—"Waterproof Cover for Carrying Kit," NC 235/1, November 18, 21, 15, 1916.
Balfour, Captain Eustace. *Drill of a Cyclist-Infantry Section*, no publication details, 1889.
—and Major Lloyd. *Drill of a Cyclist-Infantry Section*, no publication details, 1897.
—"Military Cycling, After Mr. H.G. Wells," *Fortnightly Review*, Vol. 75, February, 1901, pp. 294-303.
Barrow, J. "Departmental Bicycles - Average Cost of Repair," report submitted to the Chief Inspector, Stores and Transport, Postmaster-General's Department, Melbourne, by the Superintendent of Stores and Transport, August 6, 1940.
Beake, Lieutenant David. "Report on Dunlop Relay Ride. Easter 1912," A & I Staff, Australian Archives, 1912.
Bean, C. E. W. *Anzac to Amiens: A Shorter History of the Australian Fighting Services in the First World War*, Canberra: Australian War Memorial, 1946.
—Official Historian, Diaries, Nos. 44, 49, 1916; No. 103, 1917; No. 218, 1918.

—*Photographic Record of the War, The Official History of Australia in the War of 1914-1918*, Vol. XII, Sydney: Angus and Robertson Ltd., 1937.

—*The Australian Imperial Force in France, 1916, The Official History of Australia in the War of 1914-1918*, Vol. III, Sydney: Angus and Robertson Ltd., 1938.

Bethell, Nicholas. *Russia Besieged*, World War II, Alexandria, Virginia: Time-Life Books, 1977.

Bliss, R. "Report of BGZ. R. Bliss," *Annual Report*, U.S. Secretary of War, Vol. 1, 1896, p. 169.

Blumenson, Martin. *Liberation*, World War II, Alexandria, Virginia: Time-Life Books, 1978.

Bond, Brian; and Ian Roy, eds.. *War And Society: A Yearbook of Military History*, London: Croom Helm, 1975.

Bond, E. C. Personal letter, Pulborough, Sussex, in the Imperial War Museum, recounting his experiences in the City of London Cyclist Company, February 15, 1963.

Boudarel, Georges. "Introduction," in Giap, General Vo Nguyen, *Banner of People's War, the Party's Military Line*, London: Pall Mall Press, 1970, pp. xi-xxvi.

Bowden, Tim. *Changi Photographer: George Aspinall's Record of Captivity*, Sydney: Australian Broadcasting Corporation, 1984.

Breytenbach, J. H. *Kommandant Danie Theron*, publisher unknown, 1950.

—*Die Geskiedenis van die Tweede Vryheidsoorlog in Suid-Afrika, 1899-1902*, Deel IV, Die Boereterugtog vit Kaapland, Pretoria: Die Staatsdrukker, 1977.

Brice, Martin H. *Stronghold: A History of Military Architecture*, London: B. T. Batsford Ltd., 1984.

Bridges, Colonel W. T. "Report by Col. W. T. Bridges on Administration and Functions of the General Staff," CA 6, Australian Department of Defence [I], Commonwealth Record Series A289, 1909.

—Letter to the Secretary, the Dunlop Rubber Company of Australasia, Ltd., May 17, 1909.

Brown, Anthony Cave. *Bodyguard of Lies*, London: W. H. Allen, 1976.

Brunswig, Hans., Personal correspondence, Hamburg Germany, August 12, 1993.

Bruce, Major I. R. C. G. "Cyclist Units," *Journal of the Royal United Service Institution*, Vol. 80, 1935, pp. 641-642.

Burchett, Wilfred G. *Vietnam: Inside Story of the Guerilla War*, New York: International Publishers, 1965.

Burckart, Julius. "Das Rad im Dienste der Wehrkraft," in von Salvisburg, Paul, *Der Radfahrsport in Bild und Wort*, Munchen, 1897 (facsimile edition, New York: Hildesheim, 1980), pp. 137-156.

Butlin, S. J. *War Economy 1939-1942*, Canberra: Australian War Memorial, 1955.

"By Officers." *Regimental History of New Zealand Cyclist Corps in the Great War 1914-1918*, Auckland: Whitcomb & Tombs Ltd., 1922.

"By An R.A.A.F. Officer." *Great Was the Fall: A Story of the Malayan Tragedy*, Perth [Western Australia]: Paterson's Printing Press Ltd., 1945.

Cabell, Lieutenant De R. C. "Military Bicycling Thro" the Dakotas," *Outing*, June, 1896, pp. 214-220.

Caidin, Martin; and Jay Barbree. *Bicycles in War*, New York: Hawthorn Books, 1974.

Callahan, Raymond. *The Worst Disaster: The Fall of Singapore*, Newark: University of Delaware Press, 1977.

Cannell, Kathleen. "Parisians Returf Air Raid Trenches: Bicycles and New Variety of Uniforms Mark Change in Scene on Boulevards," *New York Times*, July 14, 1940, p. 18.

Caunter, C. F. *The History and Development of Cycles: As Illustrated by the Collection of Cycles in the Science Museum*, Part I, Historical Survey, London: HMSO, 1955.

Chapman, F. Spencer. *The Jungle is Neutral*, London: Corgi Books, 1974.

Charlton, Peter. *War Against Japan 1941-1942*, Australians at War Series, Sydney: John Ferguson and Time-Life Books, 1988.

Chief of the General Staff (Australia). "Training of Regimental Signallers," Defence files 1984/1/43.

Chongkhadikij, Theh. "The Ho Chi Minh Highway," *Bangkok Post, Sunday Magazine*, August 25, 1974, pp. 14-15.

Churchill, Winston S. *My African Journey*, London: Hodder & Stoughton, 1908.

—*The Hinge of Fate*. The Second World War, Vol. IV, London: Cassell & Co., Ltd., 1951.

Clark, A. S. (editor). *The London Cyclist Battalion: A Chronicle of Events Connected with the 26th Middlesex (Cyclist) V.R.C., and the 25th (C. of L.) Cyclist Battalion, The London Regiment, and Military Cycling in General*, London: Forster, Groom & Co., Ltd., 1932.

Collier, Richard. *D-Day, 6th June, 1944, The Normandy Landings*, New York, St. Abbeville Press, 1990.

Collis, Maurice. *Last and First in Burma*, London: Faber, 1956.

Command and General Staff School [U.S.]. "German Tactics of Combating Guerrillas," *Military Review*, Vol. 24, No. 3, June, 1944; digested from the *Soviet Handbook of the Guerrilla*, 3rd ed., 1942.

Cooper, Matthew. *The German Army, 1933-1945: Its Political and Military Failure*, London: Macdonald and Jane's, 1978.

Crisswell, Colin. *Far East History 1870-1952*, Hong Kong: Longman, 1978.

Croizat, V. J. *Lessons of the War in Indochina*, Santa Monica: The Rand Corporation, 1967.

Croker, F. P. U. "The Man-Powered Military Vehicle," *The Army Quarterly and Defence Journal*, Vol. 101, No. 4, July 1971, pp. 475-478.

Croon, Ludwig. "Das Fahrrad und seine Entwicklung," Abhandlungen und Berichte: Deutsches Museum, 1939.

Crowe, Major J. H. V. *Handbook of the Armies of Sweden and Norway*, London: HMSO, 1901.

Cunningham, Lieutenant A. J. Personal diary, [Australian War Memorial], 1914-1918.

Daley, Lieutenant Colonel Ron Reid. *Selous Scouts: Top Secret War*, Alberton, South Africa: Galago Publishing (Pty), 1982.

Davidson, Major R. P. "Some Recent Experiments in Military Bicycling," *Army and Navy Journal*, August 7, 1897, p. 902.

Davitt, Michael. *The Boer Fight for Freedom*, New York, 1902.

Deighton, Len. *Blitzkrieg*, London: Triad/Granada, 1980.

Delmotte. *Etude Sommaire de Velocipidie*, Liege, 1897.

Denis, C. *Conference sur la Velocipede Militaire*, Bordeaux: A La Librairie Nouvelle, 1888.

Denny, Harold. "Normandy Battle is Hedge to Hedge," *New York Times*, 6 July 1944, p. 5.

Department of the Army. "Counterguerilla Operations," *Department of the Army Field Manual FM 31-16*, February, 1963.

—"Military Vehicles," *Department of the Army Technical Manual TM9-2800*, October, 1947.

—"U.S. Army Counterinsurgency Forces," *Department of the Army Field Manual FM 31-22*, November, 1963.

Dilley, Roy. *Japanese Army Uniforms and Equipment: 1939-1945*, London: Almark Publishing Co., 1970.

Dollar, Charles M. "Putting the Army on Wheels: The Story of the Twenty-Fifth Infantry Bicycle Corps," *Prologue, Journal of the National Archives*, Vol. 17, No. 1, Spring, 1985, pp. 6-23.

Downey, Fairfax. "It Wasn"t Always Boxcars," *The American Legion Weekly*, September 18, 1925, pp. 10-11, 20-21.

Doyle, A. Conan. *The Great Boer War*, London: George Bell & Sons, 1900.

Duncan, H. O. *The World on Wheels*, 2 Vols., Paris: H.O. Duncan, 1926.

Duranty, Walter. "Russian Athletes Having a New day," *New York Times*, February 14, 1937, p. 37.

Eddleman, C. D. "Self-Propelled Doughboys," *Infantry Journal*, Vol. 45, pp. 234-235, 238-239, date unknown.

Egan, Charles E. "Sale of Bicycles to Adults Curbed by Order of WPB," *New York Times*, April 3, 1942, p. 1.

Elliot-Bateman, Michael. *Defeat in the East*, London: Oxford University Press, 1967.

Ellis, John. *A Social History of the Machine Gun*, London: Croom Helm, 1975.

—*Eye Deep in Hell*, London: Croom Helm, 1976.

Ellis, John T. *The Airborne Command and Center: Army Ground Forces Study No. 25*, Washington D.C.: A.G.F. Historical Section, 1946.

Elson, Robert T. *Prelude to War*, World War II, Alexandria, Virginia: Time-Life Books, 1976.

England, H. H. *War-Time Cycle Lamp & Lighting Regulations. How to make "Cycling's" lamp mask*, English Universities Press, 1943.

English Army. *Instructions for the Use of Permanent Staff Instructors of the Supplementary Reserve, Territorial Army and O.T.C., in Regard to the Care, Inspection and Repair, etc., of Small Arms, Machine Guns, and Bicycles*, 1933.

—*Handbook of the Military-Pedal-Bicycle*, 1933.

Falk, Stanley. *Seventy Days to Singapore: The Malayan Campaign, 1941-1942*, London: Robert Hale, 1975.

Fall, Bernard B. *Hell in a Very Small Place: The Siege of Dien Bien Phu*, London: Pall Mall Press, 1967.

—*Street Without Joy*, London: Pall Mall Press, 1964.

—*The Viet-Minh Regime: Government and Administration in the Democratic Republic of Vietnam*, New York: Institute of Pacific Relations, 1956.

Field, L. M. *The Forgotten War: Australian Involvement in the South African Conflict of 1899-1902*, Melbourne: University of Melbourne Press, 1979.

Fitzpatrick, Jim. "Anzacs at War on Bicycles," *Royal Historical Society of Victoria Journal*, Vol. 54, No. 3, 1983, pp. 31-38.

—"Arthur Richardson," *Australian Dictionary of Biography*, Melbourne: Melbourne University Press, Vol. 11, p. 379.

—"The Bicycle and the Australian Military - 1890-1918: A Study of the Perception and Use of Technology," Second Military History Conference, Australian War Memorial, Canberra, 1982.

—*The Bicycle and the Bush: Man and Machine in Rural Australia*, Melbourne: Oxford University Press, 1980.

—"The Bicycle in Rural Australia: A Study of Man, Machine and Milieu," Ph.D. dissertation, Human Sciences Program, Australian National University, 1979.

—"The Bicycle and the Australian Military: 1890-1918," *Hemisphere*, Vol. 27, No. 6, May/June, 1983, pp. 341-346.

—"The Early Development of Australian Road Maps," *The Globe, Journal of the Australian Map Curators" Circle*, No. 13, pp. 13-29.

—*Major Taylor in Australia*, Star Hill, 2011.

—"War on Wheels: Major Jack Hindaugh, Commander of the 1st Anzac Cyclist Battalion," *This Australia*, Vol. 2, No. 1, Summer 1982-83, pp. 34-38.

Fletcher, Marvin E. "The Black Bicycle Corps," *Arizona and the West*, Vol. 16, No. 3 (Autumn), pp. 219-232, date unknown.

—"The Negro Soldier and the United States Army, 1891-1917," Ph.D. Dissertation, University of Wisconsin, MHRC E185.63F54, 1968, pp. 64-68 on cycles.

Foot, M. R. D. *S O E in France: An Account of the Work of the British Special Operations Executive in France, 1940-1944*, Frederick, Maryland: University Publications of America, Inc., 1966.

Forster, Colin. *Industrial Development in Australia*, Canberra: Australian National University Press, 1964.

Freeth, Vincent. *The Bruce Small Story*, Melbourne: Vincent H. Freeth, 1945.

Furstenberg, Oberleutnant A. D. *Radfahrfibel Zusammengestellt und Bearbeitet*, Berlin: Verlag "Offene Worte," 1941(?).

Gammage, Bill. *The Broken Years: Australian Soldiers in the Great War*, Ringwood, Victoria: Penguin Books, 1975.

General Staff [U.K.]. Cyclist Training (Provisional), London: HMSO, 1914; 1917.

—*Handbook of the Swiss Army*, London: HMSO, 1911.

Gerard, H. *Le Probleme de l'"Infanterie montee resolu par l'"emploi de la bicyclette*, Paris: Librairie Militaire L. Baudein, 1894.

—*Infanterie Cycliste in Campagne*, Paris: Berger-Levrault et Cie, 1898.

Giap, General Vo Nguyen. *Banner of People's War, the Party's Military Line*, London: Pall Mall Press, 1970.

—*Dien Bien Phu*, Hanoi: Foreign Languages Publishing House, 1959.

Giddings, Major Howard A. "The Bicycle in the Army," *Harper's Weekly*, date unknown, p. 364.

—*Manual for Cyclists for the Use of the Regular Army, Organized Militia, and Volunteer Troops of the United States*, Kansas City, Mo.: Hudson-Kimberley Publishing Co., 1898.

Griffin, Harry H. *Bicycles and Tricycles of the Year 1886*, London: L. Upcott Gill, 1886. Reprinted by Olicana Books Limited, 1971.

Grunberger, Richard. *A Social History of the Third Reich*, Ringwood, Victoria: Penguin Books, 1974.

Guiges, Claude. "Logistique Vietnam," *Indochine-Sudest Asiatique*, March 1953.

Gullett, H. S. *Sinai and Palestine: The Official History of Australia in the War of 1914-1918*, Vol. VII, Sydney: Angus & Robertson, 4th ed., 1937.

Hammond, L. Report submitted to the Superintendent of Stores and Transport, Postmaster-General's Department, Melbourne, by the Senior Cycle Mechanic, July 26, 1940.

Harbutt, Major Noel Cambridge. "An African Experience," *Bath Cycling Club Gazette*, February 1911; reprinted in *Africana Notes and News*, Vol. 21, No. 8, December, 1975, pp. 337-339.

Harmond, Richard. "Progress and Flight: An Interpretation of the American Cycle Craze of the 1890s," *Journal of Social History*, Vol. 5, 1971, pp. 235-257.

Harrison, A. E. "The Competitiveness of the British Cycle Industry," *Economic History Review*, Vol. 22, No. 2, 1969, pp. 287-303.

Herr, Michael. *Dispatches*, New York: Alfred A. Knopf, 1977.

Hill, Lieutenant R. G. "The Capabilities and Limitations of the Bicycle as a Military Machine," *Journal of the U.S. Military Service Institutions*, July-November, 1895, pp. 312-322.

Hillier, G. Lacy. "The Use of the Cycle for Military Purposes," *Longman's Magazine*, July, 1887, pp. 268-275.

Hindhaugh, Major Jack. Personal diaries and letters, 1914-1918, made available courtesy of the late Cecily Adams, Castlecrag, N.S.W., Australia.

Holderness, G. W. "Cyclists in Salonika: Life on Struma Front," *The Mosquito*, No. 129, March, 1960, pp. 3-7.

Hoare, Robert. *World War Two: An Illustrated History*, London: Macdonald and Company, 1973.

Hobart, Major F. W. A. *Pictorial History of the Machine Gun*, London: Ian Allan, 1971.

Howton, Hugh. "Revolutionary War Machine," *Soldier*, Vol. 24, No. 4, April, 1968, pp. 43-46.

Hutchison, Lieutenant-Colonel S. S. *Machine Guns: Their History and Tactical Employment*, London: Macmillan and Co., Limited, 1938.

Irving, David. *The War Between the Generals*, London: Allen Lane, 1981.

James, Harry. "Looking Back," *Dunlop Gazette*, pp. 6-7, 12-13, August, 1939.

Joss, Lieutenant Wynyard. Personal letter to Mr. C. Hopkins, May 29, 1901, made available courtesy of R. L. Wallace.

Kaitsevaede Staap (Estonia). *Kaitsevae Jalgratta Kasiraamat*, Tallin: Kaitsevaede Staabi vi Osakonna Valjanne, 1932.

Karnow, Stanley. *Vietnam: A History*, New York: Viking Press, 1983.

Keegan, John. *The Second World War*, Sydney:

Hutchinson Australia, 1989.

Kirby, Major-General S. Woodburn. *The Loss of Singapore, The War Against Japan*, Vol. 1, London: HMSO, 1957.

Kitchener of Khartoum, Field Marshal Viscount. *Memorandum on the Defence of Australia*, Melbourne: Government Printer, 1910.

Kohn, R. S. *Bicycle Troops*, Columbus, Ohio: Remote Area Conflict Information Center, Battelle Memorial Institute. Prepared under contract SD-171 for the Advanced Research Projects Agency, Office of the Secretary of Defense, Washington, D.C., 1965.

—Personal interview, April 5, 1983.

Komer, Ambassador Robert. "Tactics and Technology," in Thompson, W. Scott, and Donaldson D. Frizzell, eds. *The Lessons of Vietnam*, St. Lucia, Queensland: University of Queensland Press, 1977, pp. 173-182.

Kolko, Gabriel. *Anatomy of a War: Vietnam, the United States, and the Modern Historical Experience*, New York: Pantheon Books, 1985.

Krigsministeriet (Denmark). *Cykelreglement for Haeren*, Kjobenhavn: Trykt, 1910.

—*Instruktion for Cyclens Brug i Haeren*, Kjobenhavn: Centraltrykkeriet, 1900.

Kruger, Rayne. *Good-bye Dolly Gray: The Story of the Boer War*, London: Cassell, 1959. Paperback edition as *Goodbye Dolly Gray: The Story of the Boer War*, London: Pan Books, 1983.

Lacouture, Jean. "Preface," in Giap, General Vo Nguyen, *Banner of People's War, the Party's Military Line*, London: Pall Mall Press, 1970, pp. vii-x.

Ladd, James D. *Inside the Commandos: A Pictorial History from World War Two to the Present*, London: Arms and Armour Press, 1984.

Laffin, John. *Damn the Dardanelles*, Sydney: Doubleday, 1980.

Lawton, First Lieutenant E. P. "The Bicycle in Military Use," *Journal of the U.S. Military Service Institutions*, Vol. 21, 1897, pp. 449-461.

Le Herisse, R. *Le Cyclisme Militaire*, Paris: Henri Charles-Lavauzelle, 1897.

Levy, Bert. "Street Fighting," *Infantry Journal*, September, 1942, pp. 23-29.

Liddell Hart, Basil, ed. *History of the Second World War*, New York: G. P. Putnam's Sons, 1970.

Liddell Hart, B. H. *The Tanks: The History of the Royal Tank Regiment*, London: Cassell, 1959.

Liebers, Arthur. "More Bicycles to Keep Nation on Wheels," *New York Times*, March 22, 1942, Section IX, p. 1.

Lloyd, J. Barclay. *One Thousand Miles with the C.I.V.* [City Imperial Volunteers], London: Methuen & co., 1901.

Lockhart, Greg. *Nation in Arms: The Origins of the People's Army of Vietnam*, Sydney: Allen & Unwin, 1989.

—Personal interview, August 23, 1993.

Long, Gavin. *The Six Years War: A Concise History of Australia in the 1939-45 War*, Canberra: Australian War Memorial and Australian Government Publishing Service, 1973.

Lunn, Hugh. *Vietnam: A Reporter's War*, St. Lucia: University of Queensland Press, 1985.

Macartney, Major John M. "Portable Bicycle for Mobile Infantry," *The United Service Magazine*, Vol. XIV, pp. 388-395.

Mackinnon, Major-General W. H. *The Journal of the C.I.V. in South Africa*, London: John Murray, 1901.

Maclear, Michael. Vietnam: *The Ten Thousand Day War*, London: Thames Methuen, 1981.

McGonagle, Seamus. *The Bicycle in Life, Love, War and Literature*, London: Pelham, 1968.

McKernan, Michael. *All In!: Australia During the Second World War*, Melbourne: Nelson, 1983.

—*Australians in Wartime: Commentary and Documents*, Melbourne: Nelson, 1980.

McNeill, Ian. "Australian Army Advisers: Perceptions of Enemies and Allies," in Maddock, Kenneth; and Barry Wright, *War: Australia and Vietnam*, Sydney: Harper and Row Publishers, 1983, pp. 37-66.

McWhirter, Norris. *Guinness Book of Records*, London: Guinness Superlatives Limited, 1977.

Manchester, William. *American Caesar: Douglas MacArthur,*

1880-1964, Melbourne: Hutchinson, 1978.

—*The Glory and the Dream: A Narrative History of America. 1932-1972*, Boston: Little, Brown and Company, 1973.

Mansfield, Sue (interviewed by Sam Keen). "War as the Ultimate Therapy," *Psychology Today*, Vol. 16, No. 6, June, 1982, pp. 56-66.

Maree, D. R. "Bicycles in the Anglo-Boer War of 1899-1902," *Military History Journal*, Vol. 4, No. 1, June 1977, pp. 15-21; republished as *Bicycles During the Boer War: 1899-1902*, Pretoria: South African National Museum of Military History, 1977.

Marr, David. Personal interview, January 12, 1979.

Martin, Ernest J. "The Cyclist Battalions and their Badges, 1888-1921," *Journal of the Society for Army Historical Research*, Vol. 22, No. 91, Autumn, 1944, pp. 277-280.

Mather, Alfred W. B. "Cyclists for the Transvaal," *Western Mail* (Perth, Western Australia), 24 Febraury 1900, p. 19.

—"Proposed Cyclists" Corps," *Morning Herald* (Perth, Western Australia), February 16, 1900, p. 2.

Maurice, Lieutenant Colonel F. *Sir Frederick Maurice: A Record of his Work and Opinions, with Eight Essays on Discipline and National Efficiency*, London: Edward Arnold, 1913.

Maurice, Major-General Sir Frederick. "Cycling and Home Defence," *The Times*, 24 May, 1901, p. 3.

—*History of the War in South Africa*, London: Hurst and Blackett Limited, 1907.

May, Lieutenant William T. *Cyclists" Drill Regulations, United States Army*, Boston: Pope Manufacturing Co., 1892.

—"On the March," 1892, an instruction and drill manual, no details of publication.

Mayo, Lida. *Bloody Buna*, Canberra: Australian National University Press, 1975.

Meintjies Johannes. *The Anglo-Boer War 1899-1902: A Pictorial History*.

Miles, General Nelson A. "Military Cycling," *Scientific American*, January 28, 1893, p. 54.

Miller, Don. "The Handlebar Infantry," *Army*, Vol. 30, September, 1980, pp. 38-40, 43.

Miller, Forbes. *Australia Since the Camera: The Second World War*, Melbourne: F.W. Cheshire, 1971.

Miller, Russell. *The Resistance, World War II*, Alexandria, Virginia: Time-Life Books, 1979.

Miller, Thomas S. *Bicycle Tactics for the Instruction, Exercise and Maneuver of Bicycle Organisations*, Chicago: T. S. Miller, no date.

Ministere del la Guerre (Belgium). *Instruction pour les Compagnies Cyclistes*, Bruxelles: Guyet Freres, 1906.

Ministerio della Guerra (Italy). *Istruzione sul Materiale Ciclistico*, Roma: Istituto Poligrafico dello Stato Libreria, 1927; 1937.

—*Istruzione sull"affardellamento dei Bersaglieri Ciclisti e Motociclisti e sul Trasporto della Armi e delle Munizione sulla Bicicletta e sui Motomezzi*, Roma: Istituto Poligrafico dello Stato Libreria, 1939.

Mitchell, W.C. *A Code of Signals for Bicycles*, Denver: Marsh and Carter, 1895.

Moseley, Leonard. *The Battle of Britain*, World War II, Alexandria, Virginia: Time-Life Books, 1979.

Moss, Second Lieutenant James A. "Bicycle Corps, 25th Inf.," *Army and Navy Journal*, September 5, 1896, p. 11.

—"The Bicycle for Military Purposes," Army and Navy Register, August 29, 1896, pp. 130-131.

—"Organization of a Bicycle Corps in Havana, Cuba," letter to the Adjutant General, October 25, 1898.

—"Recent Experiments in Infantry Bicycling Corps," *Outing*, February, 1897, pp. 488-492.

—"25th Infantry Bicycle Corps," *Army and Navy Journal*, October 2, 1897, pp. 71-72.

—"25th Infantry Bicycle Corps," *Army and Navy Journal*, July 3, 1897, p. 814.

—"25th U.S. Infantry Bicycle Corps," *Army and Navy Journal*, July 31, 1897, p. 887.

—"25th U.S. Infantry Bicycle Corps," *Army and

Navy Journal, August 7, 1897, p. 903.

Murray, Lieutenant Colonel P. L., ed. *Official Records of the Australian Military Contingents to the War in South Africa*,

Nakada, Tudao. *Imperial Japanese Army and Navy Uniforms and Equipment*, London: Arms and Armour Press, 1975.

Nalder, Major-General R. F. H. *The Royal Corps of Signals: A History of its Antecedents and Development*, London: Royal Signals Institution, 1958.

Nixon, Lieutenant Edwin. "Report on Military Aspect of Dunlop Relay Ride from Adelaide to Sydney, Carried Out During Easter 1912," A & I Staff, May 13, 1912.

O"Ballance, Edgar. *The Indo-China War (1945-1954): A Study in Guerilla Warfare*, London: Faber and Faber, 1964.

O"Dea, Eamon. "When Connecticut Put the Cavalry on Bicycles," *Yankee*, August, 1979, pp. 68-71.

O"Farrell, M. "A Deed that Saved an Empire," *The Austral Wheel*, Vol. 1, No. 12, December, 1896, pp. 353-354; and Vol. 2, No. 1, January 1897, pp. 11-14.

—"Our Delay Ride," *The Austral Wheel*, October, 1896, pp. 299-300.

Oliver, Smith Hempstone; and Donald Berkebile. *Wheels and Wheeling: The Smithsonian Institution Cycle Collection*, Washington, D.C.: Smithsonian Institution Press, 1974.

Olson, James S, ed. *Dictionary of the Vietnam War*, New York: Greenwood Press, 1988.

O"Neill, Robert J. *The German Army and the Nazi Party, 1933-1939*, New York: James H. Heineman, Inc., 1966.

Opperman, Hubert. *Pedals, Politics and People*, Sydney: Haldane Publishing, 1977.

Ordway, Albert. *Cycle-Infantry Drill Regulations of the District of Columbia National Guard*, Washington, D.C.: Judd and Detweiler, 1892.

Pakenham, Elizabeth. *Jameson's Raid*, London: Weidenfeld and Nicolson, 1960.

Pakenham, Thomas. *The Boer War*, London: Weidenfeld and Nicolson, 1979.

—Personal interview, March, 1983.

Palmer, Arthur Judson. *Riding High: The Story of the Bicycle*, London: Vision Press, 1958.

Parliament of the Commonwealth of Australia. *Annual Report*, Tariff Board, Canberra: AGPS, 1936, 1937, 1942.

Parritt, Lieutenant Colonel B. A. H. *The Intelligencers: The Story of British Military Intelligence Up to 1914*, no place, no date.

Pemberton, A. C. *The Complete Cyclist*, London: A.D. Innes & Co., 1897.

Perrett, Bryan. *A History of Blitzkrieg*, London: Robert Hale, 1983.

Perry, P. J. "Working-Class Isolation and Mobility in Rural Dorset, 1837-1936: a Study of Marriage Distances," *Institute of British Geographers, Transactions*, No. 46, March, 1969, pp. 121-141.

Pethebridge, S. A. Letter from the Acting Secretary, Department of Defence, to Dunlop Rubber Co. Ltd., February 25, 1909.

Pierre, Francis. "L"apparition de La Velocipedie Militaire," *Gazette des Armes*, No. 3, October, 1975, pp. 24-29.

Pike, Douglas. *PAVN: People's Army of Vietnam*, Novato California, Presidio Press, 1986.

Pollock, Wilfrid. *War and A Wheel: The Graeco-Turkish War as Seen from a Bicycle*, London: Chatto & Windus, 1897.

Postmaster-General (Australia). "Departmental Bicycles - Renewal of," memo sent from the Chief Inspector, Stores and Transport, Postmaster-General's Department, Melbourne, to the Director-General, Melbourne, May 21, 1935.

—"Memo" sent from the Chief Inspector, Stores and Transport, Postmaster-General's Department, Melbourne, to the Deputy-Director, Melbourne, November 27, 1935.

—"Departmental Bicycles - Average Cost of Repair," letter sent from the Chief Inspector, Stores and Transport, Postmaster-General's Department, Melbourne, to the Superintendent, Stores and Transport, all states,

April 24, 1940.

Pretorius, General Philip. Director, South African National Museum of Military History, personal correspondence, January 20, 1993.

Prior, Lieutenant-Colonel B. H. L. *Cyclist Infantry: Some Comments and Suggestions Relative to Their Work*, London: Good Ltd., undated.

—*The Military Cyclist: Notes*, Norwich: F. H. Goose, 1907.

Proctor, W. J. Letter to Captain R. H. M. Collins, C. M. G., Secretary for Defence, Victoria Barracks, Melbourne, January 20, 1909.

Puttkamer, Gerhard Frhr. v. "Die Fahrschule," in von Salvisburg, Paul, *Der Radfahrsport in Bild & Wort*, Munchen, 1897 (facsimile edition, New York: Hildesheim, 1980), pp. 49-55.

Quigley, Colonel Hugh E. "Employment of Bicycles," *Letter Report of Evaluation, Project No. 1B-158*, Army Concept team n Vietnam, APO San Francisco, JRATA, August 11, 1965.

Reischauer, Edwin O.; and Albert M. Craig. *Japan: Tradition and Transformation*, Boston: Houghton Mifflin Company, 1974.

Reichskriegministerium. *Das Truppenfahrrad*, Berlin: Mittler & Gohn, 1936.

Remington, Frederic. "Stirrups or Pedals? The Colonel of the First Cycle Infantry," *American West*, Vol. XI, No. 5, September, 1975, pp. 10-13, 58-59.

Reynolds, Captain E. H. Memo on transport costs, Australian Department of Defence file 1862/3/207, February 23, 1912.

Richardson, Michael. "Wheel to Wheel in Hanoi," *Sydney Morning Herald*, January 5, 1974, p. 8.

Richter, Klaus. *Weapons and Equipment of the German Cavalry, 1935-1945*, Atglen, PA: Schiffer Military History, 1995, originally published as *Waffen und Ausrustung der Deutschen Kavallerie* 1935-1945, Dorheim: Podzun-Pallas-Verlag.

Riley, Edward. Personal interview, Collie, Western Australia, December 11, 1976.

Ritchie, Andrew. *King of the Road: An Illustrated History of Cycling*, Berkeley: Ten Speed Press, 1975.

Robinson, Commander C. N., ed. *The Transvaal War Album*, London: Geo. Newnes, Ltd., circa 1902.

Rompel, Frederick. *Heroes of the Boer War*.

Roy, Jules. *The Battle of Dienbienphu*, London: Faber and Faber, 1965.

Royal Commission on the War in South Africa, 4 Vols. London: HMSO.

Rubenstein, David. "Cycling Eighty Years Ago," *History Today*, August, 1978, pp. 544-547.

Russell, George. "We Can Move Anywhere," *Time*, March 15, 1982, pp. 6-7.

RVNAF. *War Material Used by Vietcong in South Vietnam, Handbook II*, J2, High Command, undated.

Saito, Yoshiki. *Moshin Mare Shingaporu*, Tokyo: Gakken, no date.

Salisbury, Harrison. *Behind the Lines-Hanoi: December 23, 1966-January 7, 1967*, New York: Harper and Row, 1967.

—Personal interview, April 6, 1983.

—"North Vietnam Runs on Bicycles," *New York Times*, January 7, 1967, pp. 1, 3.

—*The 900 Days; The Siege of Leningrad*, London: Pan Books, 1971.

Salter, Stephen. "The Perils of Being Simple," *New Scientist*, February 25, 1982, pp. 495-497.

Seagrave, Sterling. *Lords of the Rim*,

Seaton, Albert. *The German Army 1933-45*, London: Wiedenfeld and Nicolson, 1983.

Sears, Roebuck & Company. *1908 Sears, Roebuck Catalogue*, Chicago: Sears, Roebuck and Company, 1908. Reprinted, Northfield, Illinois: Digest Books, Inc., 1971.

Shaul, Lieutenant E. W. "A Memory of the Advance to the Aisne," *The Britannia*, No. 8, October, 1930, pp. 51-53.

Showalter, Dennis E. "Weapons, Technology and the Military in Metternich's Germany: A Study in Stagnation?," *Australian Journal of Politics and History*, Vol. 24, No. 2, August, 1978, pp. 227-238.

Simpkins, Major B. G. *Rand Light Infantry*, Cape

Town: Howard Timmins, 1965.

Slonaker, John. "Bikin" and Hikin" Through Paradise," *Vignettes of Military History*, Vol. II, Carlisle Barracks, PA.: U.S. Army Military History Institute, No. 91, February, 20, 1978, p. 45.

Smith, Philip R., Jr. "Bicycles Built for War," *The American Legion Magazine*, November, 1977, pp. 22-23.

Smith, Robert A. *A Social History of the Bicycle: Its Early Life and Times in America*, New York: American Heritage Press, 1972.

Smith, Terence. "Cyclists Race in Vietnam with Politics in Tandem," *New York Times*, January 19, 1970, p. 3.

Snow, R. F. "The Great Bicycle Delirium," *American Heritage*, Vol. 26, June, 1975, pp. 61-72.

Soleil, F. *Etude sur la Velocipede Militaire en Belgique et a l"Etranger*, Brussels: Imprimerie O. Hoeree, 1892.

Sturmey, Henry, ed. *The Cyclist Year Book 1898*, London: Illife & Son, 1898.

Sweeting, A. J., ed. "World War II," *The Australian Encyclopaedia*, 4th edition, Sydney: The Grolier Society, 1983.

Tames, Richard. *Japan in the Twentieth Century*, London: Batsford Academic, 1981.

Tate, Stephen T., MAJ USA, *Human Powered Vehicles in Support of Light Infantry Operations*, M.A. Military Arts and Science thesis, Fort Leavenworth Kansas, AD-A211 795, 1989.

Taylor, A. J. P. *The First World War; An Illustrated History*, Middlesex, England: Penguin, 1966.

Terraine, John. "The Spectre of the Bomber," *History Today*, Vol. 32, April, 1966, pp. 4-9.

Thayer, Thomas C. "Patterns of the French and American Experience in Vietnam," in Thompson, W. Scott, and Donaldson D. Frizzell, eds. *The Lessons of Vietnam*, St. Lucia, Queensland: University of Queensland Press, 1977, pp. 17-38.

Thompson, Lieutenant Colonel P. W. "The Japanese Army, Part 2, Jungle War in Malaya," *Infantry Journal*, May, 1943, pp. 10-45.

Thompson, W. Scott, and Donaldson D. Frizzell, eds. *The Lessons of Vietnam*, St. Lucia, Queensland: University of Queensland Press, 1977.

Thorne, Captain Joseph, "Army Cycles," *Journal of the Royal United Service Institution*, Vol. 80, 1935, pp. 642-643.

Toland, John. *The Rising Sun: The Decline and Fall of the Japanese Empire, 1936-1945*, New York: Random House, 1970.

—Personal interview, April 5, 1983.

Tran Do, *Recits sur Dien Bien Phu*, Hanoi: Editions En Langues Etrangeres, 1962.

Trapman, Captain A. H. "The Cycle in Warfare: Its Potency as a Strategical and Tactical Factor," *The Times*, December 17, 1908, p. 9.

Travers, T. H. E. "Future Warfare: H.G. Wells and British Military Theory, 1895-1916," in Bond, Brian; and Ian Roy, eds., *War and Society: A Yearbook of Military History*, London: Croom Helm, pp. 67-87, 19....

Tritton, H. P. "Duke." *Time Means Tucker*, Sydney: Shakespeare Head Press, 1964.

Truong Nhu Tang. *Journal of a Vietcong*, London: Jonathan Cape, 1986.

Tsuji, Colonel Masanobu. *Singapore: The Japanese Version*, Sydney: Ure Smith, 1960.

Tuchman, Barbara, *The Guns of August*, New York: Bantam Books, 1978.

Turley, William S. *The Second Indochina War: A Short Political and Military History, 1954-1975*, Boulder, Colorado: Westview Press, 1986.

Turner, Charles. "Military Cycling," *Outing*, December, 1890, pp. 189-194.

"U. K.." "The Bicycle in the Bavarian Army," *The Chautaquan*, February, 1897, pp. 567-568, translated from *Ueber Land Und Meer* (no other details given).

U.S. Army Concept Team in Vietnam. "Employment of Bicycles," mimeo letter, APO San Francisco, August 11, 1965.

Uyeda, Teijiro; and Hiroshi Koyasu. *Small-scale Industries of Japan: The Bicycle Industry*, Japanese Council Papers, No. 8, Tokyo: Japanese Council, Institute of Pacific Relations, 1936.

Walker, Martin. "An Ersatz Army," *History Today*, June, 1982, pp. 59-60.

Wallace, R. L. *The Australians at the Boer War*, Canberra: The Australian War Memorial and the Australian Government Publishing Service, 1976.

—"Queenslanders at the Front, Boer War," personal manuscript made available to the author, April 26, 1981.

Walther, Herbert. *The Waffen SS: A Pictorial History*, West Chester PA: Schiffer Publishing, 1990, originally published as *Die Waffen SS*, Dorheim: Podzun-Pallas-Verlag.

War Department (U. S.). "Technical Manual TM-E 30-480," *Handbook on Japanese Military Forces*, Washington: U.S. Government Printing Office.

—"Technical Manual TM-E 30-451," *Handbook on German Military Forces*, Washington: U.S. Government Printing Office, 1945, reprinted by Lousiana State University Press, Baton Rouge, 1990.

War History Office. *Mare Shinko Sakusen* [Malay Offensive Operation], Tokyo: War History Office, Defense Training Institute, 1969.

War Office. *Cyclist Drill*, London: HMSO, 1900.

—*Cyclist Training*, London: HMSO, 1902; 1908; 1914; 1917.

—*Handbook on Military Bicycles*, London: HMSO, 1911; amended, 1914; 1922.

—*Handbook of the Military (Pedal) Bicycle*, London: HMSO, 1933.

—*Instruction for Armourers ... and for the Care of Bicycles*, London: HMSO, 1904.

—*Instruction for the Care and Preservation of Bicycles*, London: HMSO, 1904; 1905.

—*Manual of Military Law*, London: HMSO, 1914.

—*Trumpet and Bugle Sounds for the Army, 1914*, London: HMSO, 1914.

Watson, Roderick; and Martin Gray. *The Penguin Book of the Bicycle*, Harmondsworth: Penguin Books, 1978.

Weeks, P. A. "Departmental Bicycles - Average Cost of Repairs," report submitted to the Chief Inspector, Stores and Transport, Postmaster-General's Department, Melbourne, July 16, 1940.

Wells, H. G. "The Cyclist Soldier," *Fortnightly Review*, Vol. 74, December, 1900, pp. 914-928.

—"The Land Ironclads," *The Strand Magazine*, 1903, reprinted in 1916; my citations from *The War in the Air and Other War Forebodings*, New York: Charles Scribner's Sons, 1926, pp. 383-415.

—"The Soldier Cyclist," *Fortnightly Review*, Vol. 75, 1901, pp. 572-574.

—*The Wheels of Chance*, London, 1898.

White, Brudenell. Letter to Major Buckley, War Office, London, Australian Department of Defence File (unnumbered), August 30, 1911.

Whiting, Charles. *The Home Front: Germany*, Alexandria, Virginia, Time-Life Books, 1982.

Whitney, Lieutenant Henry H. "The Adaptation of the Bicycle to Military Uses," *Journal of the U.S. Military Service Institutions*, Vol. 17, July-November, 1895, pp. 542-563.

Whitt, Frank Rowland; and David Gordon Wilson. *Bicycling Science: Ergonomics and Mechanics*, Cambridge: M.I.T. Press, 1974.

Wigmore, Lionel. *The Japanese Thrust, Australia in the War of 1939-1945*, Series One, Army, Canberra: Australian War Memorial, 1957.

Williams, J. R. "To Battle on Bikes," *Sabretache, Journal of the Military Historical Society of Australia*, Vol. 19, No. 4, October, 1978, pp. 26-29.

—"Push Bike Soldiers," *Firm & Forester*, Vol. 5, No. 2, November, 1978, pp. 123-125.

Wilson, David Gordon. "Human Muscle Power in History," in McCullagh, James C., ed., *Pedal Power*, Emmaus, Penn.: Rodale Press, pp. 1 - 36.

Wilson, H. W. *With the Flag to Pretoria: A History of the Boer War of 1899-1900*, 2 Vols., London:

Harmsworth Brothers, 1901.

—*After Pretoria: The Guerilla War. The Supplement to "With the Flag to Pretoria,"* 2 Vols., London: Harmsworth Brothers, 1902.

Wilson, Lt. J. Cook. *A Manual of Cyclist Drill for the Use of The Cyclist Section of the Oxford University Rifle Volunteer Corps*, London: Simpkins, Marshall and Co., and Oxford: B.H. Blackwell, 1889.

Wilson, S. S. "Bicycle Technology," *Scientific American*, March, 1973, pp. 81-91.

Wood, Col. W. A., Jr. "Bicycles for Jungle Warfare," memorandum to the Chief of Ordnance, Headquarters, Army Service Forces, 1942.

Yass, Marion, *The Home Front: Britain, 1939-45*, East Sussex: Wayland, 1971.

Yergin, Daniel. *The Prize: The Epic Quest for Oil, Money, and Power*, New York: Simon & Schuster, 1991.

Younger, R. M. "Bicycles Have a New Importance and Popularity in War-Time," *Adelaide Advertiser* (South Australia), February 7, 1942, p. 11.

Zich, Arthur. *The Rising Sun*, World War II, Alexandria, Virginia: Time-Life Books, 1977.

Index

A. G. Spalding, 25
Advanced Research Projects Agency, 195, 212
Afghanistan, 205, 213
Albemarle, Earl of, 16
Allen, R.L.K., 115
Ambrose, Stephen, 135
American Expeditionary Forces (World War I), 97
Ammer, Max, 74
Angola, 208
Anzac Cyclist Battalions (World War I), 100-107
Army Cyclist Corps (Great Britain), 113
Aspinall, George, 152-153
Australia
    bicycle manufacturing, 166-170
    fuel rationing, 166
    World War II, 149-150
Australian cyclists
    in Boer War, 70-72
    Dunlop Military Dispatch Cycle Rides and, 79-83
    in World War I, 100-107
Balfour, Eustace, 88-90
Balkan Wars, 86
Barclay, J. Lloyd, 66, 207
Battelle Memorial Institute, Remote Area Conflict Information Center, 195-196
Battle of La Cateau (World War I), 96
Battle of Somme (World War I), 98
Battlefield tactics, 43-45
Beaded-edge tires, 10
Belgian cyclists
    post-World War, 83
    19th-century,
    World War I, 94, 96, 101, 115
Belgium, 122
Bersaglieri (World War I), 107-109
Bicycle clinometer, 39-40

Bicycle Corps, The: *America's Black Army on Wheels*, 205.
Bicycle manufacturing
    19th-century, 10
    Australia, 166-170
    Great Britain, 164
    Japan, 142-143
    Vietnam, 201
    Bicycle relays, 18-19
Bicycle Troops (Battelle Memorial Institute), 195
Bicycles, *See also* Military cycling;
    benefits of, 13-14
    civilian reliance during World War II on, 157-158
    cost of, 10
    folding, 3-39, 134-135, 211-213
    heavy weapons mounted on, 31-36
    military adoption of, 45-50
    military attitudes toward, 45-50
    military value of, 1-2
    ordinary, 4-5
    origin of term, 3
    overview of, 1-2
    as personal belongings, 95
    used with tanks, 87, 120-121
Bicycles and Tricycles of the year 1886 (Griffin), 5
Bicycling, concept, 1
Birdwood, General, 103
Boer War
    bicycle spies in, 72-75
    Boer use, 64-65
    British cyclists in, 65-70
    colonial cyclists in, 70-72
    Kitchener's strategy for cyclists in, 75-76
    military operations of, 54-59
    overview of, 51-52
    war cycles in, 59-63

Bombings
    Vietnam War, 200-201
    World War I, 105
    World War II, 158
Bond, E. C., 91
Boneshaker, 3
Boer War
    Boers and bicycles, 64
    British experience, 65-70
    colonial cyclists, 70
    cyclist numbers, 75
    Marshall Law, 72
    railway war cycles, 59
Bridges, Sir William T., 79, 80, 103
British cyclists, *See also* Great Britain; *specific wars*
    in Boer War, 65-70 *See also* Boer War
    post-World War I views of, 113-114
    19th-century, 14, 16, 45-46
    turn-of-the-century, 77-78
    in World War I, 94-100
    in World War II, 133-139
British Expeditionary Force (World War I), 91
Broadbent, George, 40
Bruce, Major, 115
Buller, General, 55
Bullet damage, 16
Burchett, Wilfred, 190-193, 196
Burma, 146, 152, 155
Cabell, DeR. C, 21
Callahan, Raymond, 150
Canadian cyclists, 72, 135
Cannons, 31
Cavalry, 14-17, 44
Celliers, J. D., 54
Chamberlain, Joseph, 54
Chapman, F. Spencer, 149-156
Charlton, Peter, 207-208
Chinese, Tsuji and execution of, 153
Churchill, Winston, 111, 122, 139
    on Malayan Campaign, 144
City Imperial Volunteers (C.I.V.) 66-70, 207
Clark, General Wesley, 213
Clinometer, 39

Colt machine gun, cycle-mounted, 32-33, 35
Columbia bicycle, 16
Communication, with cyclists, 85
Connaught, Duke of, 46
Connecticut National Guard Signal Corps, 14, 15, 18, 43, 77, 84
Croker, F. P. U., 209
Cycle artillery, 31-36
Cycle racing, Vietnam, 203-204
Cycling manuals, 83-85
Cyclist Drill manual, 88-90
Cyclist Training manuals, 84
Daly, Ron, 210
Daufresne, Raoul, 96
Davidson, R. P., 31
Deterding, Henri, 111
Diamond frame models
    description of, 6-10
    strength of, 8
Die Radfahrtruppe, 111, 115-122
Diem, Ngo Dinh, 186
Dilley, Roy, 152
Doole, Claire, 205
Doyle, Arthur Conan, 85
Drais, Karl von, 2
Draisienne, 3-4
Dunlop, John Boyd, 10-12
Dunlop Military Dispatch Cycle Rides, 79-83
Dursley-Pedersen folding bicycle, 38
Dutch cyclists, 93, 112-115, 160
Dwarf safeties, 6
Dwyer Folding Bicycle Company, 38
Eadie hub, 9
"Easy-Rider" bicycles, Vietnam, 202
88th Airborne Infantry Battalion (U.S.), 133
Eisenhower, Dwight D., 111, 138-139, 174
El Salvador, 209
Fall, Bernard, 177, 184, 187
Finns, 128
First Indochina War, 177-179. *See also* Vietnam War
Fixed wheel machines, 8-9, 24, 36, 60-61
Folding bicycles

British in World War II, 134-136
Captain Gerard model, 36-39, 95
German paratroopers in World War II, 165
Montague model, 211-213
Foot, M. R. D., 163
France, *See also specific wars*
    civilian reliance on, World War II, 158-160
    development of draisienne in, 5-6
    resistance in, 160-164
    Vietnam and, 165-167, 172-174
    in World War II, 118, 122
Franco-Prussian War, 6, 48
Freewheel hubs, 9
Fuel shortages/rationing
    in Australia, 166
    in France, 159
    in Great Britain, 164
    synthetic fuel production and, 122
    in United States, 171
    during World War I, 111-112
Fulbright, William, 175, 194-195
Fuller, J. F. C., 211
Fusilier Battalion (World War II), 131
Gallipoli, 97, 101
Gammage, Bill, 91, 107
Geneva Accords of 1954, 186
Gerard, H., 36-37
German cyclists, *See also specific wars*
    19th-century, 14, 47
    in World War II, 115-133
Germany
    civilian reliance on, 157-158
    military cycling study, 115-122
    occupation of Rhineland by, 121
    tank use and, 124-126
Giap, Vo Nguyen, 177-179
Giddings, Howard A., 77, 84
Goldwater, Barry, 208-209
Greco-Turkish War, 42
Grandin, General, 16
Great Britain, *See also* British cyclists
    in Boer War, 65-70
    civilian reliance on bicycles, 164
    in World War I, 94-100
    in World War II, 133-139
Griffin, Harry, 5
Haig, Alexander, 208
Halberstam, David, 187
Hale, Colonel, 14, 209
Hardaway, B. E, 195
Hart, Basil Liddell, 139
Heinrichs, Kimberly, 205
Helicopters, 17, 188-189, 200-203
Herr, Michael, 188
High-wheelers, 4-6
Hill, R. G., 20
Hillier, G. Lacy, 36
Hindhaugh, Jack, 101-107
Hitler Youth Organization, 158
Ho Chi Minh, 177, 179, 189, 190, 192
Ho Chi Minh Trail, 197-201
Holderness, G. W., 96
Horsemen
    military cyclists vs., 14-17
    speed of, 7-8
    World War II, 122-133
Howard, General, 18
Humber safety bicycle, 6-7
Huy, Chu, 198
Illinois National Guard, 31
Imperial Japanese Army and Navy Uniforms and Equipment (Nakada), 207
India, 96, 98-100, 143
Infantry on bicycles
    advantages of, 15-17
    capabilities of, 19-21
    disadvantages of, 20-21
    experiments involving, 23-24
Instructions for the Care and Preservation of Military Bicycles, 84
Ironclads, *See* Tanks
Italian cyclists
    19th-century, 48
    in World War I, 107-109
Italy, 48, 85, 114
Jameson, Leander S., 52-54
Japan
    bicycle industry in, 142-143

Bombings
    Vietnam War, 200-201
    World War I, 105
    World War II, 158
Bond, E. C., 91
Boneshaker, 3
Boer War
    Boers and bicycles, 64
    British experience, 65-70
    colonial cyclists, 70
    cyclist numbers, 75
    Marshall Law, 72
    railway war cycles, 59
Bridges, Sir William T., 79, 80, 103
British cyclists, *See also* Great Britain; *specific wars*
    in Boer War, 65-70 *See also* Boer War
    post-World War I views of, 113-114
    19th-century, 14, 16, 45-46
    turn-of-the-century, 77-78
    in World War I, 94-100
    in World War II, 133-139
British Expeditionary Force (World War I), 91
Broadbent, George, 40
Bruce, Major, 115
Buller, General, 55
Bullet damage, 16
Burchett, Wilfred, 190-193, 196
Burma, 146, 152, 155
Cabell, DeR. C, 21
Callahan, Raymond, 150
Canadian cyclists, 72, 135
Cannons, 31
Cavalry, 14-17, 44
Celliers, J. D., 54
Chamberlain, Joseph, 54
Chapman, F. Spencer, 149-156
Charlton, Peter, 207-208
Chinese, Tsuji and execution of, 153
Churchill, Winston, 111, 122, 139
    on Malayan Campaign, 144
City Imperial Volunteers (C.I.V.) 66-70, 207
Clark, General Wesley, 213
Clinometer, 39

Colt machine gun, cycle-mounted, 32-33, 35
Columbia bicycle, 16
Communication, with cyclists, 85
Connaught, Duke of, 46
Connecticut National Guard Signal Corps, 14, 15, 18, 43, 77, 84
Croker, F. P. U., 209
Cycle artillery, 31-36
Cycle racing, Vietnam, 203-204
Cycling manuals, 83-85
Cyclist Drill manual, 88-90
Cyclist Training manuals, 84
Daly, Ron, 210
Daufresne, Raoul, 96
Davidson, R. P., 31
Deterding, Henri, 111
Diamond frame models
    description of, 6-10
    strength of, 8
Die Radfahrtruppe, 111, 115-122
Diem, Ngo Dinh, 186
Dilley, Roy, 152
Doole, Claire, 205
Doyle, Arthur Conan, 85
Drais, Karl von, 2
Draisienne, 3-4
Dunlop, John Boyd, 10-12
Dunlop Military Dispatch Cycle Rides, 79-83
Dursley-Pedersen folding bicycle, 38
Dutch cyclists, 93, 112-115, 160
Dwarf safeties, 6
Dwyer Folding Bicycle Company, 38
Eadie hub, 9
"Easy-Rider" bicycles, Vietnam, 202
88th Airborne Infantry Battalion (U.S.), 133
Eisenhower, Dwight D., 111, 138-139, 174
El Salvador, 209
Fall, Bernard, 177, 184, 187
Finns, 128
First Indochina War, 177-179. *See also* Vietnam War
Fixed wheel machines, 8-9, 24, 36, 60-61
Folding bicycles

British in World War II, 134-136
Captain Gerard model, 36-39, 95
German paratroopers in World War II, 165
Montague model, 211-213
Foot, M. R. D., 163
France, *See also specific wars*
civilian reliance on, World War II, 158-160
development of draisienne in, 5-6
resistance in, 160-164
Vietnam and, 165-167, 172-174
in World War II, 118, 122
Franco-Prussian War, 6, 48
Freewheel hubs, 9
Fuel shortages/rationing
in Australia, 166
in France, 159
in Great Britain, 164
synthetic fuel production and, 122
in United States, 171
during World War I, 111-112
Fulbright, William, 175, 194-195
Fuller, J. F. C., 211
Fusilier Battalion (World War II), 131
Gallipoli, 97, 101
Gammage, Bill, 91, 107
Geneva Accords of 1954, 186
Gerard, H., 36-37
German cyclists, *See also specific wars*
19th-century, 14, 47
in World War II, 115-133
Germany
civilian reliance on, 157-158
military cycling study, 115-122
occupation of Rhineland by, 121
tank use and, 124-126
Giap, Vo Nguyen, 177-179
Giddings, Howard A., 77, 84
Goldwater, Barry, 208-209
Greco-Turkish War, 42
Grandin, General, 16
Great Britain, *See also* British cyclists
in Boer War, 65-70
civilian reliance on bicycles, 164
in World War I, 94-100

in World War II, 133-139
Griffin, Harry, 5
Haig, Alexander, 208
Halberstam, David, 187
Hale, Colonel, 14, 209
Hardaway, B. E, 195
Hart, Basil Liddell, 139
Heinrichs, Kimberly, 205
Helicopters, 17, 188-189, 200-203
Herr, Michael, 188
High-wheelers, 4-6
Hill, R. G., 20
Hillier, G. Lacy, 36
Hindhaugh, Jack, 101-107
Hitler Youth Organization, 158
Ho Chi Minh, 177, 179, 189, 190, 192
Ho Chi Minh Trail, 197-201
Holderness, G. W., 96
Horsemen
military cyclists vs., 14-17
speed of, 7-8
World War II, 122-133
Howard, General, 18
Humber safety bicycle, 6-7
Huy, Chu, 198
Illinois National Guard, 31
Imperial Japanese Army and Navy Uniforms and Equipment (Nakada), 207
India, 96, 98-100, 143
Infantry on bicycles
advantages of, 15-17
capabilities of, 19-21
disadvantages of, 20-21
experiments involving, 23-24
Instructions for the Care and Preservation of Military Bicycles, 84
Ironclads, *See* Tanks
Italian cyclists
19th-century, 48
in World War I, 107-109
Italy, 48, 85, 114
Jameson, Leander S., 52-54
Japan
bicycle industry in, 142-143

Malayan Campaign and, 144-153, 207, 211
Japanese cyclists
   in Burma, 146, 152, 155
   in Malayan Campaign, 144-153
   in New Guinea, 154, 155
Joffre, General, 95
Joss, Wynyard, 72
Journal-Examiner Yellow Fever Transcontinental Relay Ride, 18-19
The Jungle is Neutral (Chapman), 148-156
K & K Supply Company of New York, 143
Kamm, Leo, 40
Karnow, Stanley, 201
Kirby, Major-General, 144
Kitchener, Lord, 56, 75-76, 79
Kohn, S., 195-196
Koyasu, Hiroshi, 142
Kruger, Paul, 56
Kruger, Rayne, 63
Laos, 177, 179, 189, 193, 197, 200
Lloyd, J. Barclay, 66-70
London, World War I bombardments, 109
London Cyclist Battalion, 35, 46, 78, 82, 100, 113
Ludendorff, Erich, 112
Lunn, Hugh, 177, 188
MacArthur, Douglas, 170
Macartney, John M., 50
Machine guns, 50-51, 70, 91, 118, 149, 173
   cycle mounted Colt, 32, 35
   on tricycles, 31, 34, 36
   in World War I, 91
Mackinnon, W. H., 207
Malayan Campaign
   analysis of, 153-156
   Australian ambush and, 149-150, 207
   Japanese cyclists in, 144-153
Malvern Star cycle factories, 168-170
Manual for Cyclists for the Use of the Regular Army, Organized Militia, and Volunteer Troops of the United States (Giddings), 84
Manual of Cycling Drill (Wilson), 84
Manual of Military Law (War Office, London), 95
Maree, D. R., 62
Massey-Harris bicycles, 71

Masters, Peter, 135, 137
Maurice, Sir Frederick, 19, 65, 75, 85
May, William T., 1, 84
Menzies, Donald, 59-62
Mercader, George, 133
Miles, Nelson A., 18, 23
Montgomery, General, 111, 139
Moss, James A., 23-31, 196
Murrow, Edward R., 171
Nakada, Tadao, 207
National Liberation Front (NLF) (Vietcong), 186
Navarre, Henri, 178
New Guinea, 154-155
New Zealand, 100, 104-105, 110
Normandy invasion, 133-136
Notes on the Belgian Army, 83
O'Dea, Pat, 7
Office of Price Administration (U. S.), 171
Operation Market Time (Vietnam War), 198
Ordinaries, Ordinary bicycles, 4-6
Ordway, Albert, 14, 84
Osmond Cycles, 54
Parades, cyclists in, 19, 31, 49
Paratroopers, 39, 165,
"Paratrooper" folding bicycle, 211-212
Patton, George, 138-139
Penny-farthings, 4
Pentagon bicycle study, 195-196
People's Army of Vietnam (PAVN), 189
Plumer, Lieutenant Colonel, 71, 72, 75
Pneumatic tires 7, 8, 10-11, 35, 60, 92
Poland, 119, 122, 157
Pollock, Wilfred, 42-43
Pope, Albert, 4, 14, 31, 32, 77, 84
Pope Manufacturing Company, 14, 31
Prince of Wales, 141
Quadricycles, 6
Queensland Imperial Bushmen's unit (Boer War), 70
Radfahrtruppe, See *Die Radfahrtruppe*
Railway cycles, 49, 59-63
Railway lines, South African, 56-59
Rearguard, World War I cyclists in, 96, 120
Remote Area Conflict Information Center (Battelle Memorial Institute), 195

Repulse, 141
Rhodes, Cecil, 53
Rhodesia, 53, 71-72, 110, 210-211
Richardson, Arthur, 71
Richardson, Michael, 202
Road maps, 40, 80
Roberts, Lord 46, 56, 64
Robinson, C. N., 51
Rose, J. G., 61
Rowlands, E. M., 54
Roy, Jules, 196
Royal Australian Cycle Corps, 62
Royal Marine Commandos (World War II), 134
Rubber tires, 4-5, 31, 60,
Rudge cycles, 202
Rumania, 112, 122
Russia, 48, 112, 115, 122, 124, 128. *See also* Soviet Union
Russo-Japanese War, 86, 142
Safety bicycles, 6-10, *See also* Bicycles
Saito, Yoshiki, 148
Salisbury, Harrison, 124, 175, 194-196, 201-202
Saunders, Sol, 201
Savile, A. R., 14, 209
Sears Roebuck, 63
Second Company of District of Columbia National Guard, 14
Second Indochina War, 186, 190, 201. *See also* Vietnam War
2nd Battalion Tower Hamlet Volunteers, 31
Selous Scouts (Rhodesia), 210-211
Singapore, 141, 148, 150-153, 207
Singer bicycles, 46
Six twin-wheeled cycle units, 33,
Small, Bruce, 168-170
Smith, Sir Charles Holled, 45
Smith, Terence, 204
South Africa. *See* Boer War
Soviet Union. *See* Russia
Spies, 72-73
Stalin, Joseph, 122
Starley, John K., 6
Sturmey-Archer three-speed hubs, 9, 95

Sunbeam, 46
Sussex Maneuvers, 85
Suzuki, Sosaku, 153-154
Swiss cyclists, 47, 205-206
Tang, Truong Nhu, 189, 193-194, 201-202
Tanks
    development of, 86-87
    Germany and, 111
    Japanese 149, 150
    use of bicycles with, 87, 90
    in World War I, 89, 115-116
    in World War II, 111, 113, 122-126
Telegraphers, 39, 57
Territorial Force (Great Britain), 86
Theron, Danie, 64
Thiess, Rudolf, 117-121
Thieu, Nguyen Van, 204
Thompson, P. W., 148
Thompson, R. E., 41
Thorn-Proof tires, 12
Tires
    beaded-edge, 10
    development of rubber, 10-11
    pneumatic, 7-8, 10-11, 35, 60-61
Toland, John, 153
Topographers, 39
Tour de Vietnam, 203
Transvaal Cycle and Motor Corps, 78
Trench warfare, 86, 93-94, 100, 118
Tricycles
    artillery, 31, 36,
    development of, 6
    in Vietnam, 184
Tsuji, Masanobu, 141, 145, 148, 150,
    profile of, 153-154
Tuchman, Barbara, 111
Turkey, 86
Turner, Charles, 17
25th United States Infantry Bicycle Corps., 23-31
26th Middlesex Cyclist Regiment, 31, 110
Ty, Ding Van, 181
United States, *See also specific wars*
    bicycle rationing in, 171-174

cycle manuals in, 84-85
cycling in turn-of-the-century, 77
economic growth in World War II, 171
Vietnam and, 187-190
Uyeda, Teijiro, 142
Van Hoy, William, 139-140
Velocipede, 2-4, 48
Vickers Sons & Maxim, 31-33, 77
Vietnam: Inside Story of the Guerrilla War (Burchett), 190
Vietnam War
  bicycle use in, 179-186, 190-197,
  civilian bicycle use in, 201-203
  cycle racing, 203
  Dien Bien Phu, 179-186
  Geneva Accords, 186
  geographic features of, 177
  Hanoi and, 195, 201-202
  historical background of, 177-178
  Ho Chi Minh Trail and, 197-201
  overview of, 175-177
  Pentagon's bicycle study, 195-196
  propaganda and, 203-204
  technology and, 187-190
Volks Grenadier Divisions, 129
Von Brauchitsch, General, 122
War and a Wheel (Pollock), 42
War correspondents, 42-43
War cycles
  in Boer War, 51, 59-63
  design of, 59-61
  operation of, 61-63
War Production Board (U. S.), 171, 173-174
Wells, H. G., 77
  on *Cyclist Drill* manual, 88-90
  on tanks, 86-87, 116
WaveCrest, 213
West Australian Bushmen's Contingent, 71
Westmoreland, General, 174
Wheelbarrow porters, 159
Wheels of Chance (Wells), 88
White, General, 103
Whitney, Henry H., 39-40, 47
Wielrijders Rapportgangers Corps, 64, 75

Wilson, Henry, 94
Wilson, J. Cook, 84
Wilson, Lieutenant, 74
Wilson, Woodrow, 112
Wolseley, Lord, 1, 14, 209
Wolseley, Sir Evelyn, 45-46
Woodhouse, Herbert, 1

About the Author

Jim Fitzpatrick was born in Elkton, Maryland and grew up in southern California, graduating from UCLA. After serving with the Peace Corps in El Salvador, he moved to Australia, where he has spent most of his adult life and professional career. He earned a PhD in Human Sciences at The Australian National University.

He has been Director of Major Gifts for the Salvation Army in Phoenix, Arizona; Executive Administrator of the Australian Spinal Research Foundation; Project Director for a National Library of Australia Oral History Project; Research Officer with the Education Department of Western Australia; an urban planner in southern California; and has taught in the Geography Departments of the Universities of Natal, Western Australia, and New England.

He is the author of several books and numerous articles and reports on health, education, urban planning, cycling and history.

The Australian Logie Award-winning film, 'Tracks of Glory', was based on his book, *Major Taylor in Australia*.

The Bicycle and the Bush by Jim Fitzpatrick

The Bicycle and the Bush, cited in the Judges' Report of the 1981 Australian National Book Council Awards, is a superbly illustrated study of the widespread use of the bicycle in the vast Australian Outback by rural workers from 1890 into the 20th century.

'*The Bicycle and the Bush* represents the discipline of history at its best. Investigative, scholarly, pioneering, but always entertaining,' *The Review* (Australia).

'A highly satisfying piece of original social history,' *The Canberra Times*.

Major Taylor in Australia by Jim Fitzpatrick

Marshall W. (Major) Taylor, a brilliant black American cyclist of the 1890s and early 20th century, was for some years the highest paid and most famous athlete in what was then the world's most popular and lucrative sport. But he had to battle racial intolerance, dangerous and dirty tactics, and self-serving promoters, and it was in White Australia, in 1904, that it came to a head.

This is the first significant new work on Major Taylor since Andrew Ritchie's original biography in 1988, and includes an excellent collection of photographs, some never before seen in American or Australian books.

'Read your new account with great interest - splendid job!' *Andrew Ritchie, author of* Major Taylor.

'You have captured an angle of Major Taylor's life that few have touched upon,' *Karen Brown Donovan, Major Taylor's great-granddaughter.*

'Jim Fitzpatrick breaks new ground. His rigorous research and close readings of period accounts offer the reader a vivid portrayal of Taylor the man, and the notorious factors that led to Taylor quitting the sport at the top of his talents.' *Peter Nye, author of* Hearts of Lions.

'A great read … a very nice job of historical setting while keeping the narrative moving at a fast pace'. *Ronald A. Smith, Emeritus Professor of Exercise and Sport Science at the Pennsylvania State University.*

'I found it gripping and insightful and marvellously sad … a superb piece of work.' *Michael McKernan, University of New South Wales.*

www.starhillstudio.com.au

www.ingramcontent.com/pod-product-compliance
Lightning Source LLC
Chambersburg PA
CBHW080358170426
43193CB00016B/2756